Anatomy Coloring Book

for Health Professions

Notice

Medicine is an ever-changing science. As new research and clinical experience broaden our knowledge, changes in treatment and drug therapy are required. The authors and the publisher of this work have checked with sources believed to be reliable in their efforts to provide information that is complete and generally in accord with the standards accepted at the time of publication. However, in view of the possibility of human error or changes in medical sciences, neither,the authors nor the publisher nor any other party who has been involved in the preparation or publication of this work warrants that the information contained herein is in every respect accurate or complete, and they disclaim all responsibility for any errors or omissions or for the results obtained from use of the information contained in this work. Readers are encouraged to confirm the information contained herein with other sources. For example and in particular, readers are advised to check the product information sheet included in the package of each drug they plan to administer to be certain that the information contained in this work is accurate and that changes have not been made in the recommended dose or in the contraindications for administration. This recommendation is of particular importance in connection with new or infrequently used drugs.

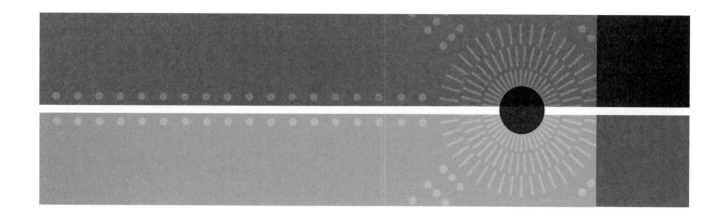

ANATOMY COLORING BOOK
FOR HEALTH PROFESSIONS

David A. Morton, PhD

Associate Professor
Anatomy Director
Department of Neurobiology and Anatomy
University of Utah School of Medicine
Salt Lake City, Utah

Medical

New York Chicago San Francisco Athens London Madrid
Mexico City New Delhi San Juan Singapore Sydney Toronto

Anatomy Coloring Book for Health Professions

1 2 3 4 5 6 7 8 9 0   QVS/QVS   18 17 16 15 14

ISBN 978-0-07-171400-6
MHID 0-07-171400-6

The book was set in Helvetica by Aptara, Inc.
The editors were Michael Weitz and Christie Naglieri.
The production supervisor was Catherine Saggese.
The text was designed by Dreamit, Inc.
The cover designer was Anthony Landi.
Quad Graphics was the printer and binder.

This book is printed on acid-free paper.

McGraw-Hill books are available at special quantity discounts to use as premiums and sales promotions, or for use in corporate training programs. To contact a representative please visit the Contact Us pages at www.mhprofessional.com.

To my father, Gordon Morton, who instilled in me a love of drawing and art…this has helped immensely in learning anatomy.

Contents

Preface

The purpose of this coloring book is to help health professional students understand the big picture of human anatomy. The vehicle used to accomplish this task is text and illustrations, which both figuratively and literally provide the "Big Picture" of human anatomy through a regional approach. The images in *Anatomy Coloring Book for Health Professionals* are based on illustrations from the textbook, *The Big Picture. Gross Anatomy*.

The format of this coloring book is simple; each page-spread consists of a list of anatomical terms on the left-hand page (even #) and their associated illustration(s) on the right-hand page (odd #). Choose a color for a structure in the left-hand column and fill in its word and reference letter. Then, find the associated letter in the image on the right-hand page and color in the structure.

Some helpful hints for your coloring experience are as follows:

- Coloring pencils work best as markers may bleed through the pages.
- The more colors you have available the more effective your complete images will look. I suggest having between 10 and 12 colors at the least, 16 to 18 is ideal.
- You are free to use any color you would like for each structure, however here is a guideline followed by most anatomy texts: orange for muscles, red for arteries, blue for veins, green for lymphatics, and yellow for nerves.
- Use light colors for large anatomical structures and dark colors for smaller structures.
- In some illustrations there are repeated identical structures, such as ribs or vertebrae. Even though each structure may lack a label you should color each of them to make your illustration complete.

White space is provided on each page for a reason. Once you have matched and colored each term with its associated structure transfer anatomy and clinical information from your anatomy lecture notes onto the appropriate coloring book page and term. In this way, you actively search, color and learn your anatomy and correlate what you have colored with what you have learned in your lectures. Your anatomy coloring book becomes your personalized anatomy textbook.

I hope you enjoy this coloring book as much as I enjoyed creating it.

—*David A. Morton, PhD*

SECTION I

BACK

1.

Back

| 1a | **Superficial Back Muscles** |

The superficial back muscles consist of the trapezius, levator scapulae, rhomboid major, rhomboid minor, and latissimus dorsi muscles. Although these muscles are located in the back, they are considered to be muscles of the upper limbs because they connect the upper limbs to the trunk and assist in upper limb movements via the scapula and humerus. The superficial back muscles receive most of their nerve supply from the ventral rami of spinal nerves (primarily the brachial plexus) and act on the upper limbs.

Color each of the following using a different color for each muscle:

A. TRAPEZIUS MUSCLE. Aids in elevation, retraction, depression, and rotation of the scapula and is innervated by the spinal accessory nerve (CN XI).

B. LEVATOR SCAPULAE MUSCLE. Elevates and rotates the scapula and is innervated by the dorsal scapular nerve (C5).

C. RHOMBOID MINOR MUSCLE. Retracts the scapula and is innervated by the dorsal scapular nerve (C5).

D. RHOMBOID MAJOR MUSCLE. Retracts the scapula and is innervated by the dorsal scapular nerve (C5).

E. LATISSIMUS DORSI MUSCLE. Adducts, extends, and medially rotates the humerus and is innervated by the thoracodorsal nerve (C6–C8).

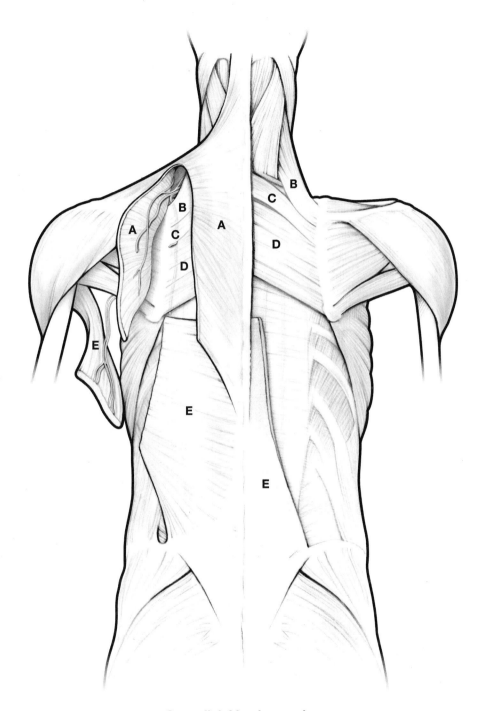

Superficial back muscles.

1b   Deep Back Muscles

The deep back muscles consist primarily of the splenius capitis, erector spinae (spinalis, longissimus, and iliocostalis), and transversospinalis (semispinalis, multifidus, and rotators) muscles. These deep back muscles are segmentally innervated by the dorsal rami of spinal nerves at each vertebral level where they attach. These muscles are responsible for maintaining posture and are in constant use during locomotion.

Color each of the following using a different color for each muscle:

A. SPLENIUS CAPITIS MUSCLE. Extends and rotates the head and neck.

Erector spinae muscles.

   B. SPINALIS MUSCLE.

   C. LONGISSIMUS MUSCLE.

   D. ILIOCOSTALIS MUSCLE.

Transversospinalis muscles.

   E. SEMISPINALIS CAPITIS MUSCLE.

   F. MULTIFIDUS MUSCLE.

   G. ROTATORES MUSCLE.

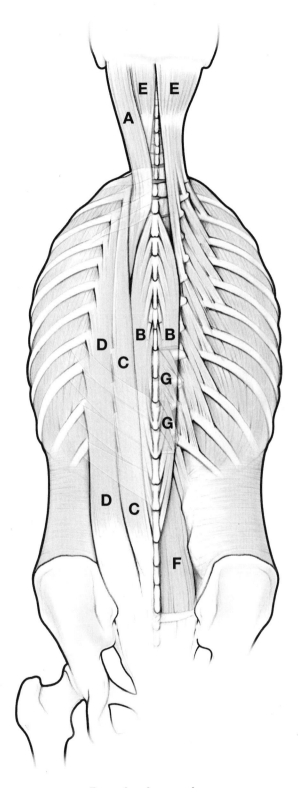

Deep back muscles.

1c   Vertebrae

The vertebral column consists of 33 vertebrae (7 cervical, 12 thoracic, 5 lumbar, 5 sacral, and 4 coccygeal). These vertebrae, along with their ligaments and intervertebral discs, form the flexible, protective, and supportive vertebral column that contains the spinal cord. Each vertebra is often referred to by its first letter to simplify its description (i.e., the third cervical vertebra is referred to as C3).

Color each of the following using a different color for each structure:

Vertebral column.

A. CERVICAL VERTEBRAE.

B. THORACIC VERTEBRAE.

C. LUMBAR VERTEBRAE.

D. SACRAL VERTEBRAE.

E. COCCYGEAL VERTEBRAE.

Color in these other bones viewed in this illustration (be sure to include both sides):

F. SCAPULA.

G. CLAVICLE.

H. RIBS.

I. OS COXA.

A typical vertebra. A typical vertebra consists of the following.

J. BODY.

K. PEDICLE.

L. TRANSVERSE PROCESS.

M. SUPERIOR ARTICULAR PROCESS.

N. INFERIOR ARTICULAR PROCESS.

O. LAMINA.

P. SPINOUS PROCESS.

Q. VERTEBRAL FORAMEN.

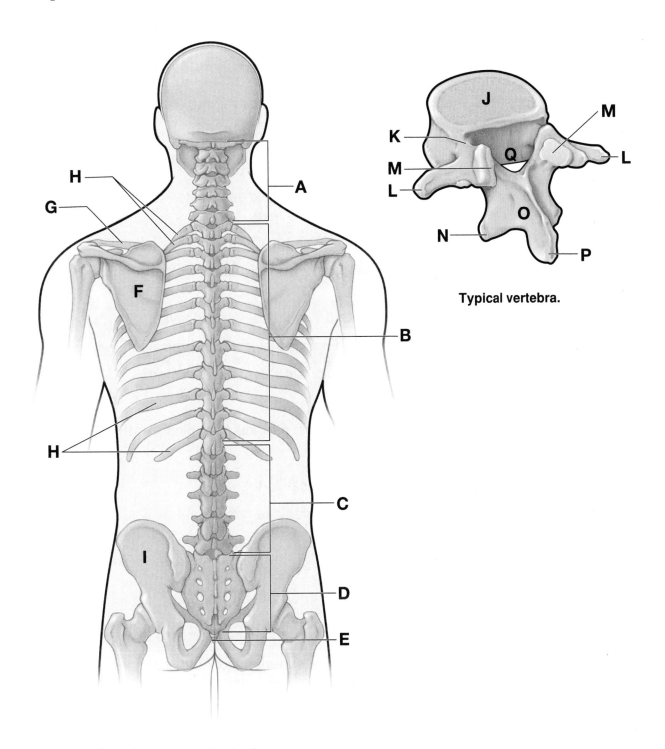

Typical vertebra.

Osteology of vertebral column.

1d — Spinal Cord, Vertebral Column, and Spinal Nerves

The spinal cord is a part of the central nervous system (CNS) that communicates with the body tissues via spinal nerves.

Color each of the following using a different color for each structure:

Spinal cord. Located within the vertebral (spinal) canal and terminates as the conus medullaris at the L1 and L2 vertebral level.

- **A.** CERVICAL SPINAL CORD. Color 8 cervical spinal cord levels.
- **B.** THORACIC SPINAL CORD. Color 12 thoracic spinal cord levels.
- **C.** LUMBAR SPINAL CORD. Color 5 lumbar spinal cord levels.
- **D.** SACRAL SPINAL CORD. Color 5 sacral spinal cord levels.
- **E.** COCCYGEAL SPINAL CORD. Color 1 coccygeal spinal cord level.

Vertebral column. The vertebral column surrounds and protects the spinal cord.

- **F.** SKULL.
- **G.** CERVICAL PEDICLES. Color 7 cervical vertebrae.
- **H.** THORACIC PEDICLES. Color 12 thoracic vertebrae.
- **I.** LUMBAR PEDICLES. Color 5 lumbar vertebrae.
- **J.** SACRAL PEDICLES. Color 5 sacral vertebrae.
- **K.** COCCYGEAL VERTEBRAE. Color 4 coccygeal vertebrae.

Spinal nerves. 31 pairs of spinal nerves formed by the dorsal and ventral roots.

- **L.** 8 CERVICAL SPINAL NERVES. Exit superiorly to 7 cervical vertebrae.
- **M.** 12 THORACIC SPINAL NERVES. Exit inferiorly to 12 thoracic vertebrae.
- **N.** 5 LUMBAR SPINAL NERVES. Exit inferiorly to the 5 lumbar vertebrae.
- **O.** 5 SACRAL SPINAL NERVES. Exit through dorsal sacral foramina.
- **P.** 1 COCCYGEAL SPINAL NERVE.

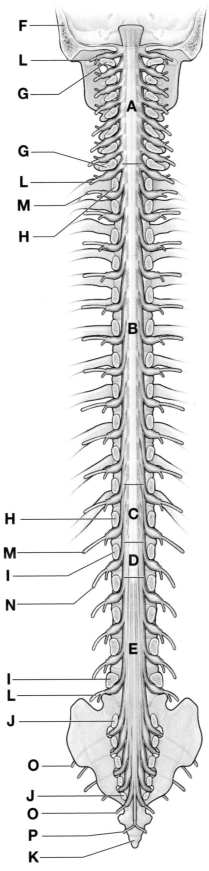

**Posterior view of the coronal section
of the vertebral canal.**

1e   Spinal Nerves and Spinal Meninges

The brain and spinal cord are protected by the vertebral column and meninges. The spinal cord communicates with the tissues of the body via spinal nerves.

Color each of the following using a different color for each structure:

Spinal meninges. Three layers of connective tissues meninges called dura mater, arachnoid mater, and pia mater protect the spinal cord.

A. VERTEBRA.

B. DURA MATER. Most superficial meninge.

C. ARACHNOID MATER. Intermediate meninge that contains CSF within the subarachnoid space.

D. PIA MATER. Deepest meninge that tethers the spinal cord to the dura mater via denticulate ligaments.

E. SPINAL CORD.

Spinal nerves. At each spinal cord segment, ventral and dorsal roots join to form spinal nerves that bifurcate into ventral and dorsal rami. These aspects of spinal nerves provide communication pathways between the spinal cord and tissues of the body.

F. DORSAL ROOT. Transports sensory neurons to the spinal cord.

G. VENTRAL ROOT. Transports motor neurons from the spinal cord.

H. SPINAL NERVE TRUNK. The union of the dorsal and ventral roots.

I. DORSAL RAMUS. Transports sensory and motor neurons to and from the back.

J. VENTRAL RAMUS. Transports sensory and motor neurons to and from the body wall and limbs.

K. SYMPATHETIC GANGLION. Houses postganglionic sympathetic neuron cell bodies.

L. GRAY AND WHITE RAMI COMMUNICANS.

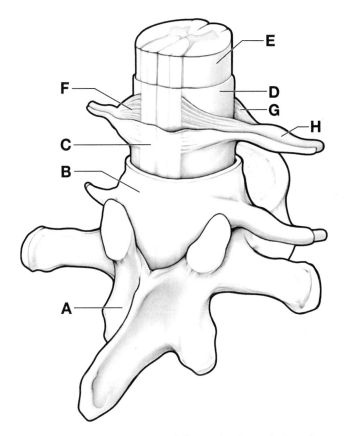

**T2 segment of the spinal cord showing
spinal meningeal layers.**

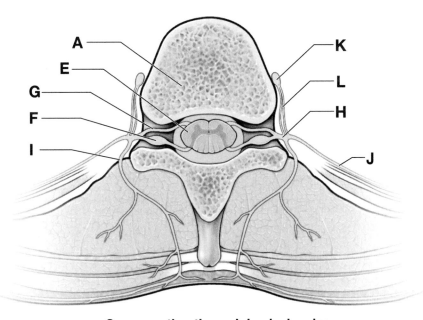

**Cross section through back showing
spinal roots, nerves, and rami.**

SECTION II

THORAX

2.

Anterior Thoracic Wall

| 2a | **Thoracic Skeleton** |

The thoracic skeleton consists of the thoracic vertebrae posteriorly, the ribs laterally, and the sternum and costal cartilages anteriorly. The costal cartilages secure the ribs to the sternum. The thoracic cage forms a protective cage around vital organs such as the heart, lungs, and great blood vessels. The thoracic skeleton provides attachment points for the muscles of the back and chest that allow support of the shoulder girdle (scapula and clavicle) and movement of the upper limbs.

A. SCAPULA.

B. THORACIC VERTEBRAE. Articulates with 12 pairs of ribs.

C. MANUBRIUM.

D. STERNAL BODY.

E. XIPHOID PROCESS.

F. TRUE RIBS. The first 7 ribs.

G. FALSE RIBS. Ribs 8, 9, and 10 articulate with sternum via costal cartilage.

H. FLOATING RIBS. Ribs 11 and 12; do not articulate with sternum.

I. COSTAL CARTILAGE. Hyaline cartilage that articulates ribs with sternum.

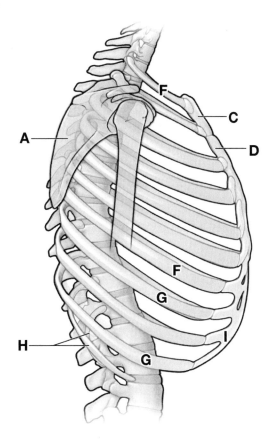

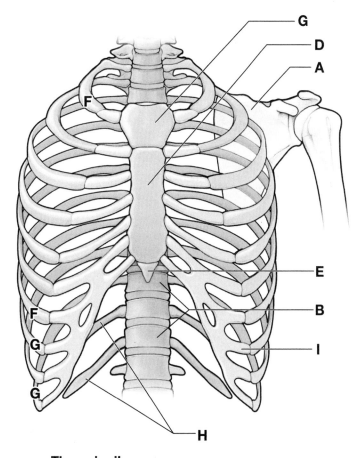

Thoracic rib cage.

2b   Breast

The breast contains the mammary gland, a subcutaneous gland that is specialized in women for the production and secretion of milk and, as a result, enlarges during the menstrual cycle and during pregnancy. Each gland consists of 15 to 20 radially aligned lobes of glandular tissue. A lactiferous duct drains each lobe. The lactiferous ducts converge and open onto the nipple, which is positioned on the anterior surface of the breast and is surrounded by a somewhat circular hyperpigmented region, the areola.

A. SKIN.

B. ADIPOSE TISSUE. Occupies the spaces between the lobes of glandular tissue.

C. RIBS. The mammary gland overlies ribs 2 through 6 between the sternum and midaxillary line.

D. LACTIFEROUS DUCT. Drain each lobe and converge to open in the nipple.

E. DEEP LAYER OF SUPERFICIAL FASCIA.

F. AREOLA.

G. SUSPENSORY LIGAMENTS. Fibrous bands that course between the dermis and superficial fascia and support the weight of the breast.

H. RETROMAMMARY SPACE. A layer of loose connective tissue that separates the breast from the deep fascia and provides some degree of breast mobility over underlying structures.

I. INTERCOSTAL MUSCLES.

J. PECTORALIS MAJOR MUSCLE.

K. DEEP PECTORAL FASCIA.

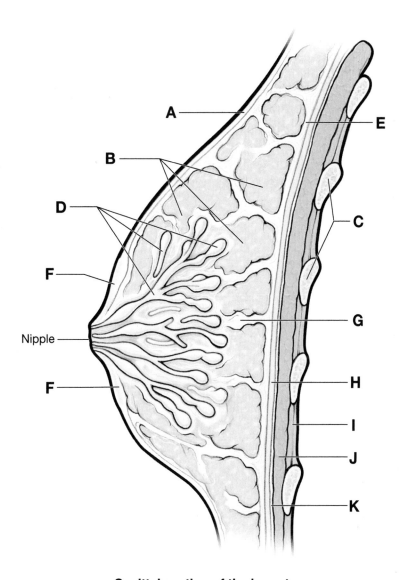

Sagittal section of the breast.

2c   Anterior Thoracic Muscles

For simplicity, we will divide muscles of the anterior thoracic wall into two groups: 1. The superficial thoracic muscles, which originate on the thoracic skeleton and inserts on the upper limb and 2. Intercostal muscles.

Superficial thoracic muscles. Attaches to the anterior portion of the thorax and primarily act on the scapula and humerus.

A. PECTORALIS MAJOR MUSCLE.

B. PECTORALIS MINOR MUSCLE.

C. SERRATUS ANTERIOR MUSCLE.

D. SUBCLAVIUS MUSCLE.

E. DELTOID MUSCLE.

F. EXTERNAL OBLIQUE MUSCLE.

G. RECTUS SHEATH.

H. LATISSIMUS DORSI MUSCLE.

I. CLAVICLE.

J. MANUBRIUM.

K. STERNAL BODY.

L. XIPHOID PROCESS.

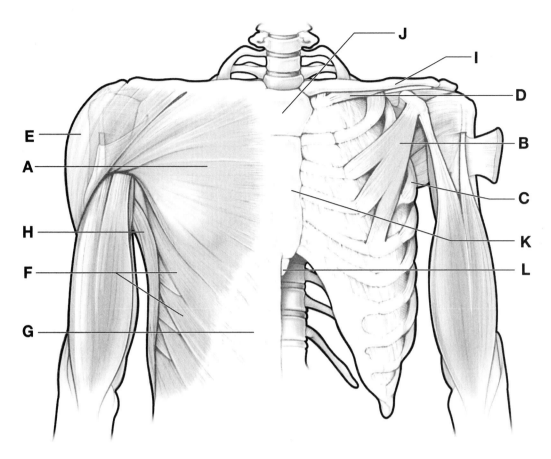

Muscles of anterior thoracic wall.

2d   Vessels and Nerves of Thoracic Wall

The nerve and blood supply of the thoracic wall consists largely of the neurovascular elements that course through the intercostal spaces. The major elements in each space consist of an intercostal vein, artery, and nerve.

Intercostal space. Consists of three layers of muscles with the intercostal vessels and nerves positioned between the internal and innermost intercostal muscles.

A. SKIN AND SUPERFICIAL FASCIA.

B. ERECTOR SPINAE MUSCLES.

C. EXTERNAL INTERCOSTAL MUSCLE.

D. INTERNAL INTERCOSTAL MUSCLE.

E. INNERMOST INTERCOSTAL MUSCLE.

F. STERNUM.

G. THORACIC VERTEBRA.

H. SCAPULA.

I. SPINAL CORD.

J. DORSAL ROOT.

K. VENTRAL ROOT.

L. INTERCOSTAL NERVE.

M. POSTERIOR CUTANEOUS NERVE.

N. AORTA.

O. POSTERIOR INTERCOSTAL ARTERY.

P. AZYGOS VEIN.

Q. POSTERIOR INTERCOSTAL VEIN.

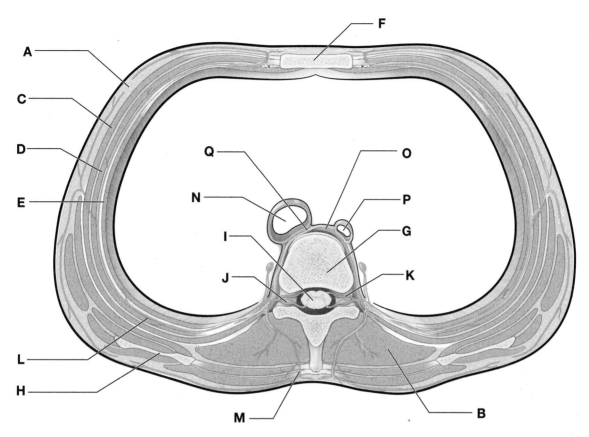

Axial section of thorax showing intercostal spaces. (Superior view)

| 2e | **Arteries of the Thoracic Wall** |

The thoracic wall receives its arterial supply from the following branches of the subclavian artery and thoracic aorta.

A. AORTA.

B. POSTERIOR INTERCOSTAL ARTERIES.

C. AORTIC ARCH.

D. BRACHIOCEPHALIC ARTERY.

E. RIGHT COMMON CAROTID ARTERY.

F. RIGHT SUBCLAVIAN ARTERY.

G. INTERNAL THORACIC ARTERIES.

H. ANTERIOR INTERCOSTAL ARTERIES.

I. ANTERIOR CUTANEOUS ARTERIES.

J. LEFT COMMON CAROTID ARTERY.

K. LEFT SUBCLAVIAN ARTERY.

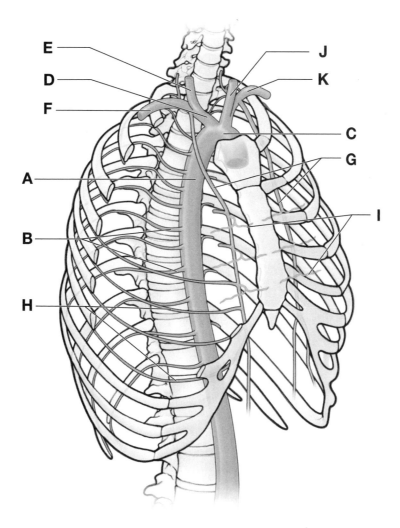

Arteries of the thoracic wall.

2f   Veins of the Thoracic Wall

Venous drainage of the thoracic wall is from the following tributaries
of the subclavian and azygos system of veins.

A. RIGHT BRACHIOCEPHALIC VEIN.

B. RIGHT SUPERIOR INTERCOSTAL VEIN.

C. POSTERIOR INTERCOSTAL VEIN.

D. AZYGOS VEIN.

E. ANTERIOR INTERCOSTAL VEIN.

F. LEFT SUPERIOR INTERCOSTAL VEIN.

G. LEFT BRACHIOCEPHALIC VEIN.

H. INTERNAL THORACIC VEIN.

I. ANTERIOR CUTANEOUS BRANCHES.

J. HEMIAZYGOS VEIN.

K. ACCESSORY HEMIAZYGOS VEIN.

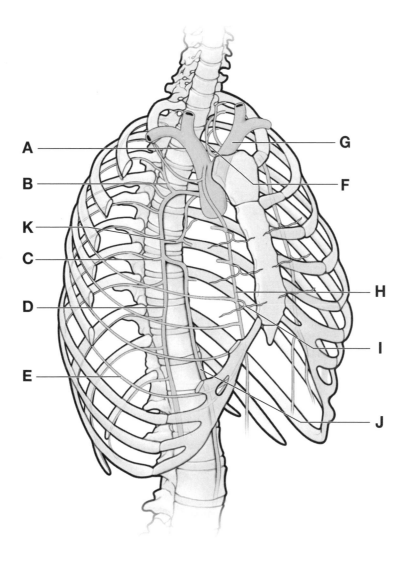

Veins of the thoracic wall.

2g   Diaphragm (Superior View)

The diaphragm is a musculotendinous structure that separates the thoracic from the abdominal cavity. The diaphragm is composed of a peripheral muscular portion and a central tendon. It is dome shaped, and upon contraction of its muscular portion, it flattens. The diaphragm is the principle muscle of respiration and is innervated by the phrenic nerve (C3–C5).

A. STERNUM.

B. MEDIASTINAL PARIETAL PLEURA.

C. COSTAL PARIETAL PLEURA.

D. DIAPHRAGMATIC PARIETAL PLEURA.

E. DIAPHRAGM.

F. ESOPHAGUS.

G. AORTA.

H. AZYGOS VEIN.

I. HEMIAZYGOS VEIN.

J. THORACIC LYMPHATIC DUCT.

K. INTERVERTEBRAL DISC.

L. SPINAL CORD.

M. THORACIC VERTEBRA.

N. RIB.

O. COSTAL CARTILAGE.

P. INTERCOSTAL MUSCLES.

Q. INFERIOR VENA CAVA.

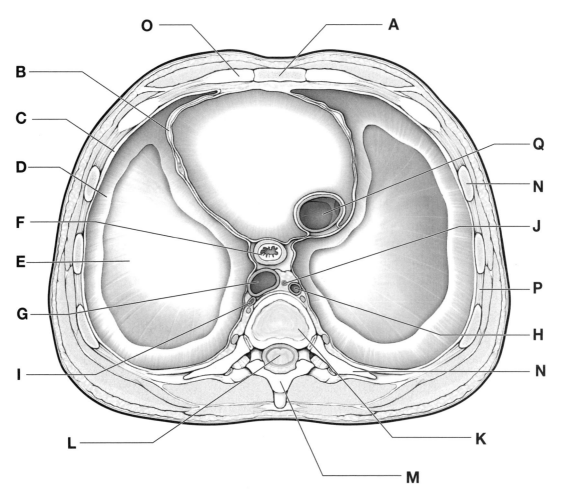

Superior view of the axial section of the thorax above the diaphragm.

3.

Lungs

| 3a | **Pleural Sacs (in situ)** |

The lungs exchange oxygen and carbon dioxide and are the functional organs of the respiratory system. To serve that vital function, the lungs are located adjacent to the heart within the pleural sacs. The pleurae are serous membranes that line the internal surface of the thoracic cavity and external surface of the lungs. The pleurae secrete fluid that decreases resistance against lung movement during breathing.

A. RIGHT BRACHIOCEPHALIC VEIN.

B. RIGHT SUBCLAVIAN VEIN.

C. RIGHT INTERNAL JUGULAR VEIN.

D. LEFT BRACHIOCEPHALIC VEIN.

E. LEFT SUBCLAVIAN VEIN.

F. LEFT INTERNAL JUGULAR VEIN.

G. BRACHIOCEPHALIC TRUNK (ARTERY).

H. LEFT COMMON CAROTID ARTERY.

I. LEFT SUBCLAVIAN ARTERY.

J. RIGHT COMMON CAROTID ARTERY.

K. RIGHT SUBCLAVIAN ARTERY.

L. AORTIC ARCH.

M. PULMONARY TRUNK.

N. PULMONARY VEINS.

O. RIGHT LUNG.

P. LEFT LUNG.

Q. DIAPHRAGM.

R. RIBS.

S. COSTAL CARTILAGE.

T. STERNUM.

U. HEART.

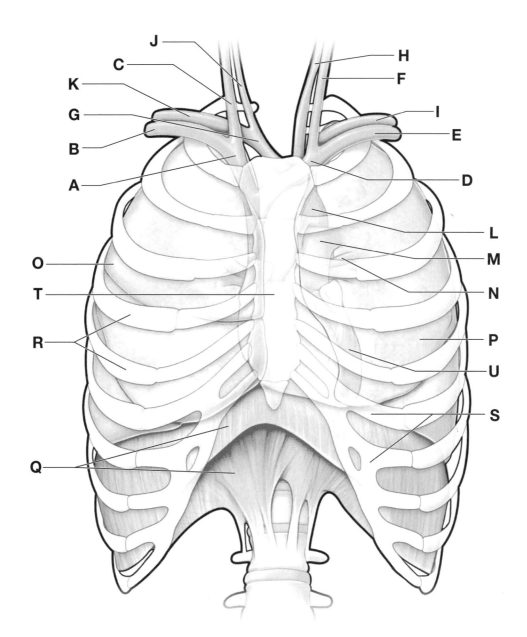

Pleural sacs.

3b   Parietal and Visceral Pleura

Parietal pleura lines the internal surface of the thoracic cavity and is assigned specific names, depending on the structures that it lines (mediastinum, costal surface, diaphragm, and cervical region). Parietal pleura receives somatic sensory innervation in contrast to visceral pleura which receives visceral sensory innervation.

A. RIBS.

B. INTERCOSTAL MUSCLES.

C. CERVICAL PARIETAL PLEURA.

D. COSTAL PARIETAL PLEURA.

E. DIAPHRAGMATIC PARIETAL PLEURA.

F. MEDIASTINAL PARIETAL PLEURA.

G. PLEURAL CAVITY.

H. COSTODIAPHRAGMATIC RECESS.

I. VISCERAL PLEURA.

J. TRACHEA.

K. RIGHT PRIMARY BRONCHUS.

L. LEFT PRIMARY BRONCHUS.

M. ESOPHAGUS.

N. AORTA.

O. AZYGOS VEIN.

P. LUNG.

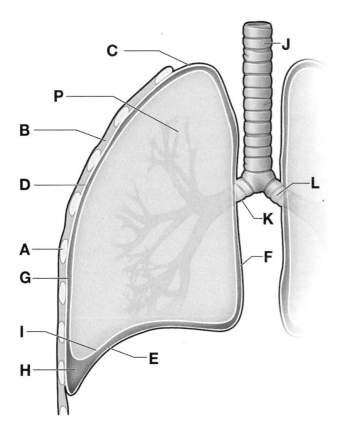

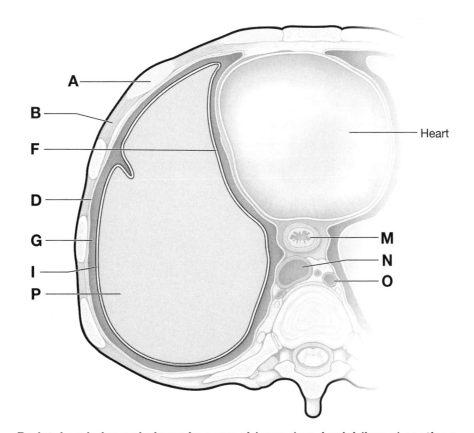

Parietal and visceral pleura in coronal (upper) and axial (lower) sections.

3c   Lungs (in situ)

The lungs are the organs of gas exchange. To serve that function efficiently, the lungs are attached to the trachea and the heart. The trachea, or windpipe, courses from the larynx into the thorax, where the trachea bifurcates into the right and left primary (principal or mainstem) bronchi at the T4–T5 vertebral level. The right primary bronchus divides into superior, middle, and inferior secondary (lobar) bronchi, corresponding to superior, middle, and inferior lobes of the right lung, respectively. The left primary bronchus divides into superior and inferior secondary bronchi, corresponding to superior and inferior lobes of the left lung, respectively. Each secondary bronchus further divides into tertiary (segmental) bronchi, which further divide. The smallest bronchi give rise to bronchioles, which terminate in alveolar sacs where the exchange of gases occurs.

A. TRACHEA.

B. RIGHT PRIMARY BRONCHUS.

C. RIGHT SECONDARY BRONCHI.

D. RIGHT TERTIARY BRONCHI.

E. LEFT PRIMARY BRONCHUS.

F. LEFT SECONDARY BRONCHI.

G. LEFT TERTIARY BRONCHI.

H. UPPER LOBE OF RIGHT LUNG.

I. MIDDLE LOBE OF RIGHT LUNG.

J. LOWER LOBE OF RIGHT LUNG.

K. UPPER LOBE OF LEFT LUNG.

L. LOWER LOBE OF LEFT LUNG.

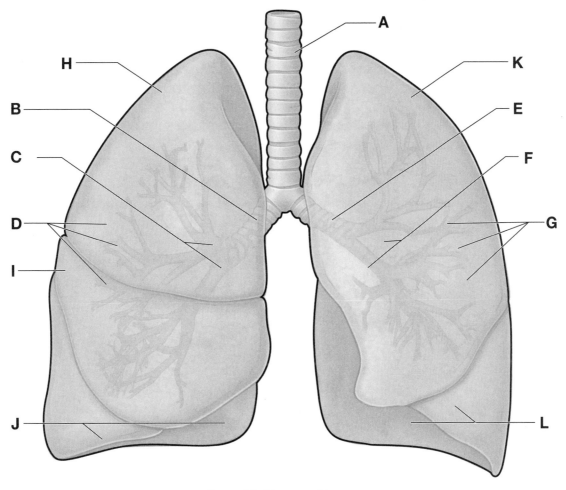

Bronchial tree and lungs.

3d   Medial of the Lungs

The right lung has three lobes (superior, middle, and inferior), which are divided by a horizontal and an oblique fissure. The right lung is shorter and wider than the left because of the higher right dome of the diaphragm and because the heart bulges more into the left side of the thorax.

A. SUPERIOR LOBE.

B. MIDDLE LOBE.

C. INFERIOR LOBE.

D. PULMONARY ARTERY.

E. PULMONARY VEIN.

F. RIGHT PRIMARY BRONCHUS.

G. PULMONARY LIGAMENT.

H. HORIZONTAL FISSURE.

I. OBLIQUE FISSURE.

The left lung has only two lobes (superior and inferior), which are divided by an oblique fissure along the sixth rib. Instead of having a middle lobe, the left lung has a space occupied by the heart. Therefore, the left lung has a cardiac notch as well as the lingula, an extension of the left superior lobe into the left costomediastinal recess.

J. SUPERIOR LOBE.

K. INFERIOR LOBE.

L. PULMONARY ARTERY.

M. PULMONARY VEIN.

N. LEFT PRIMARY BRONCHUS.

O. PULMONARY LIGAMENT.

P. OBLIQUE FISSURE.

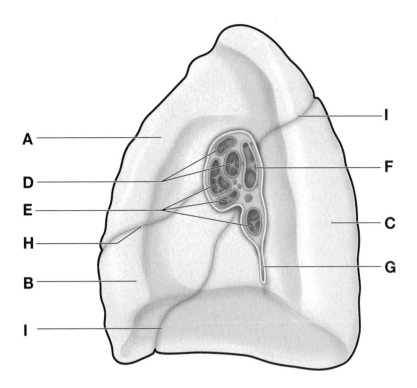

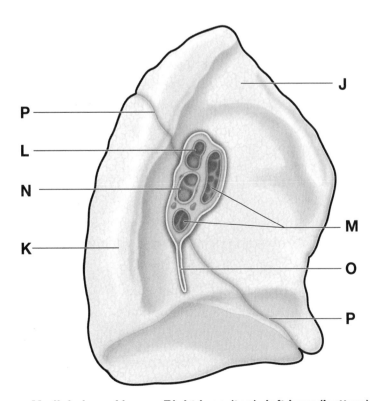

Medial view of lungs. Right lung (top); left lung (bottom).

3e   Hilum of the Lungs

For gas exchange to occur, the lung must be connected to the heart so that oxygenated blood and deoxygenated blood flow between both organs. The location where blood vessels and other structures enter and leave the lungs is called the hilum of the lung.

A. ESOPHAGUS.

B. TRACHEA.

C. RIGHT LUNG.

D. LEFT LUNG.

E. AORTIC ARCH.

F. BRACHIOCEPHALIC TRUNK.

G. LEFT COMMON CAROTID ARTERY.

H. LEFT SUBCLAVIAN ARTERY.

I. DESCENDING AORTA.

J. VAGUS NERVE (CN X).

K. LIGAMENTUM ARTERIOSUM.

L. PULMONARY TRUNK.

M. PULMONARY ARTERIES. Delivers deoxygenated blood from the systemic circulation to exchange carbon dioxide with oxygen in the lung, therefore, color these vessels blue.

N. PULMONARY VEINS. Transports oxygenated blood from the pulmonary capillaries to the left atrium of the heart in the lung, therefore, color these vessels red.

O. BRONCHI.

P. PULMONARY PLEXUS. Receives parasympathetic innervation from the vagus nerves and sympathetic innervation from T1–T4 spinal cord levels.

Q. ESOPHAGEAL PLEXUS.

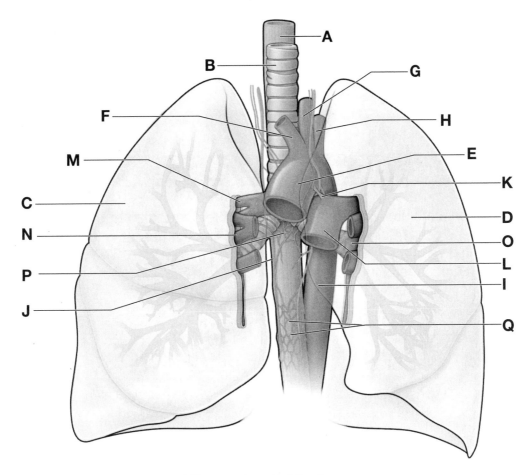

Hilum and root of the lungs.

4.

Heart

| 4a | Layers of the Heart |

The pericardium is a sac that encloses the heart, akin to the pleura that encloses the lungs. The pericardium and the heart are located in the middle of the thorax, between T4 and T8 vertebrae. The pericardium has parietal and visceral layers.

A. LUNG.

B. VISCERAL PLEURA.

C. PLEURAL SPACE.

D. PARIETAL PLEURA.

E. FIBROUS LAYER OF PARIETAL PERICARDIUM. A strong dense collagenous tissue that blends with the tunica externa of the great vessels and the central tendon of the diaphragm.

F. SEROUS LAYER OF PARIETAL PERICARDIUM. Lines the inner surface of the fibrous pericardium.

G. PERICARDIAL SPACE. Serous fluid fills this space and lubricates the beating heart.

H. EPICARDIUM. A serous layer that intimately follows the contours of the heart surface.

I. MYOCARDIUM.

J. ENDOCARDIUM.

K. LEFT VENTRICLE.

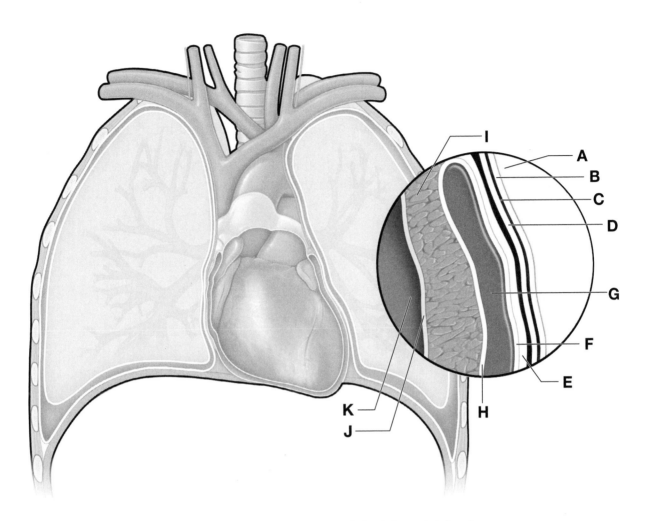

**Zoomed in coronal section of the left ventricle of
the heart demonstrating tissue layers.**

4b Superficial Aspects of the Heart

The superior border of the heart is formed by the right and left auricles of the atria. The ascending aorta and pulmonary trunk merge from this border, and the superior vena cava enters the right atrium. The right ventricle forms the anterior border. The left border is formed by the left ventricle and auricle of the left atrium. The right border is formed by the superior vena cava, right atrium, and inferior vena cava.

A. SUPERIOR VENA CAVA.

B. AORTIC ARCH.

C. PULMONARY TRUNK.

D. PULMONARY ARTERIES.

E. PULMONARY VEINS.

F. INFERIOR VENA CAVA.

G. RIGHT ATRIUM.

H. RIGHT VENTRICLE.

I. LEFT ATRIUM.

J. LEFT VENTRICLE.

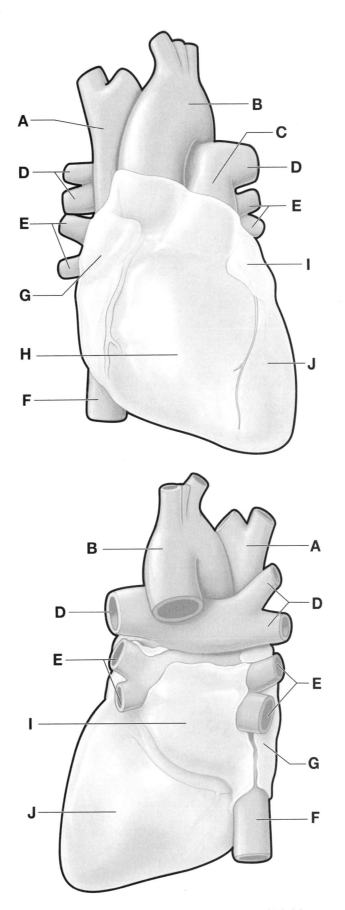

Anterior and posterior views of the superficial heart.

4c **Coronary Circulation**

Although blood fills the chambers of the heart, the myocardium is so thick that it requires its own artery–capillary–vein system, called the "coronary circulation," to deliver and remove blood to and from the myocardium. The vessels that supply oxygenated blood to the myocardium are known as coronary arteries. The vessels that remove the deoxygenated blood from the heart muscle are known as cardiac veins.

A. SUPERIOR VENA CAVA.

B. INFERIOR VENA CAVA.

C. ASCENDING AORTA.

D. LEFT CORONARY ARTERY.

E. LEFT ANTERIOR DESCENDING ARTERY.

F. LEFT CIRCUMFLEX ARTERY.

G. LEFT (OBTUSE) MARGINAL ARTERY.

H. RIGHT CORONARY ARTERY.

I. ANTERIOR CARDIAC ARTERY AND VEIN.

J. POSTERIOR INTERVENTRICULAR ARTERY.

K. CORONARY SINUS.

L. MIDDLE CARDIAC VEIN.

M. GREAT CARDIAC VEIN.

N. PULMONARY VALVE.

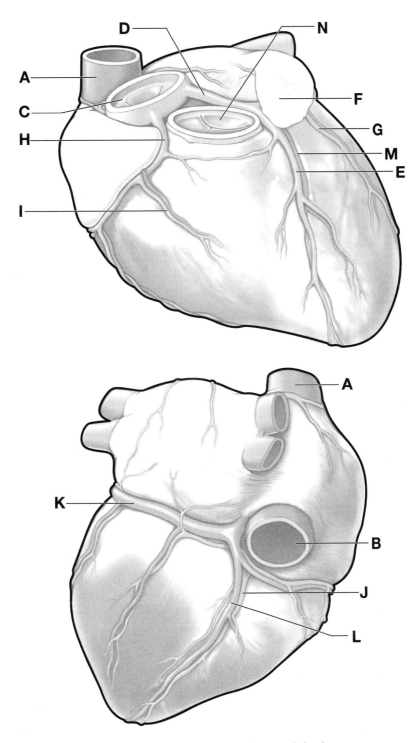

**Anterior and posterior views of the heart
showing the coronary circulation.**

4d Chambers of the Heart

The human heart is a four-chambered pump composed of cardiac muscle. There are two atria and two ventricles. The atria receive blood from organs outside the heart and pump it under low pressure to the corresponding ventricle. The two ventricles have thicker muscular walls that function to pump blood to all of the organs of the body, through capillary networks.

A. SUPERIOR VENA CAVA.

B. RIGHT ATRIUM.

C. FOSSA OVALIS.

D. TRICUSPID VALVE.

E. CORONARY SINUS.

F. INFERIOR VENA CAVA.

G. RIGHT VENTRICLE.

H. PAPILLARY MUSCLES.

I. CHORDAE TENDINEAE.

J. PULMONARY VALVE.

K. PULMONARY TRUNK.

L. LEFT ATRIUM.

M. PULMONARY VEINS.

N. MITRAL VALVE.

O. LEFT VENTRICLE.

P. AORTIC VALVE.

Q. ASCENDING AORTA.

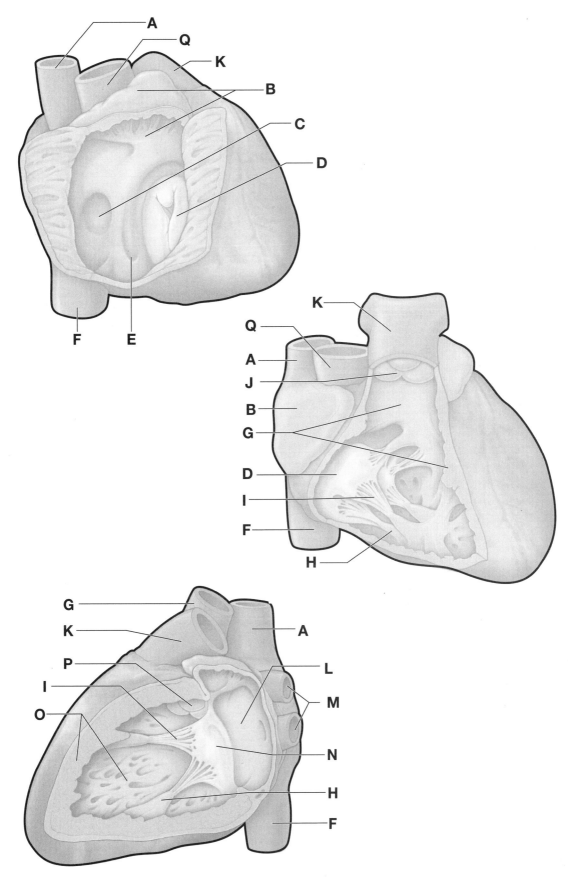

Heart chambers. Right atrium open (top); right ventricle open (middle); left atrium and ventricle open (bottom).

5.

Superior and Posterior Mediastinum

5a   Divisions of the Mediastinum

The mediastinum is the anatomic region medial to the pleural sacs bordered by the sternum, vertebral column, rib 1, and the diaphragm. The mediastinum is further divided into inferior and superior parts by a horizontal plane passing through the sternal angle to the T4–T5 intervertebral disc. The inferior mediastinum is classically subdivided into anterior, middle, and posterior parts. Therefore, the four subregions of the mediastinum are the anterior mediastinum, middle mediastinum, posterior mediastinum, and superior mediastinum.

A. TRACHEA.

B. CLAVICLE.

C. T4 VERTEBRA.

D. T5 VERTEBRA.

E. MANUBRIUM.

F. STERNAL BODY.

G. STERNAL ANGLE.

H. AORTA.

I. SUPERIOR VENA CAVA.

J. PULMONARY TRUNK.

K. ANTERIOR MEDIASTINUM.

L. MIDDLE MEDIASTINUM.

M. POSTERIOR MEDIASTINUM.

N. SUPERIOR MEDIASTINUM.

O. DIAPHRAGM.

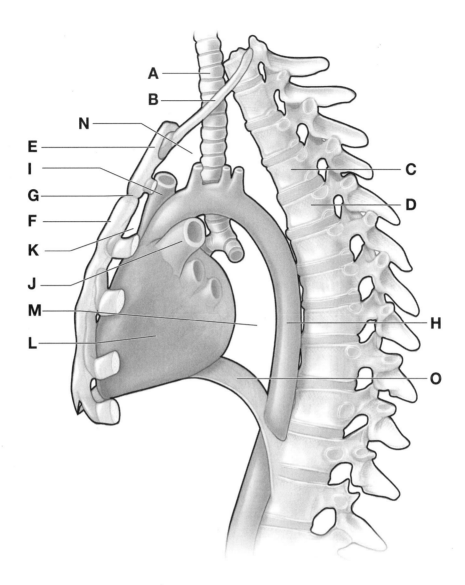

**Lateral view of the thorax illustrating the
mediastinal subdivisions.**

5b Posterior Mediastinum (Axial View)

The region containing anatomic structures deep to the pericardial sac, including the thoracic portion of the descending aorta, the azygos system of veins, the thoracic duct, the esophagus, and the vagus and sympathetic nerves.

A. SPINAL CORD.

B. THORACIC VERTEBRA.

C. ESOPHAGUS.

D. DORSAL ROOT.

E. VENTRAL ROOT.

F. DORSAL RAMUS.

G. VENTRAL RAMUS.

H. INTERCOSTAL NERVE.

I. SYMPATHETIC GANGLION.

J. THORACIC AORTA.

K. POSTERIOR INTERCOSTAL ARTERY.

L. AZYGOS VEIN.

M. POSTERIOR INTERCOSTAL VEIN.

N. HEMIAZYGOS VEIN.

O. THORACIC DUCT.

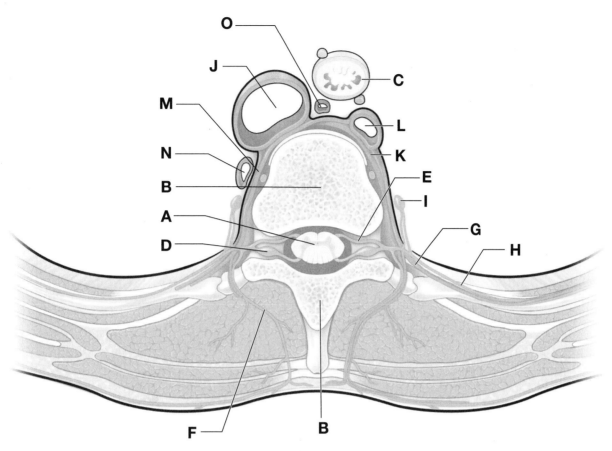

Superior view of axial section of the posterior mediastinum.

5c   Posterior Mediastinum (Anterior View)

Structures of the posterior mediastinum include the sympathetic chain, azygos system of veins (drains the posterior thoracic and abdominal walls), thoracic lymphatic duct (transports lymph), and the thoracic aorta (located left of the midline, along the anterior surface of the thoracic vertebrae).

A. ESOPHAGUS.

B. THORACIC AORTA.

C. RIGHT BRACHIOCEPHALIC VEIN.

D. LEFT BRACHIOCEPHALIC VEIN.

E. LEFT SUBCLAVIAN VEIN.

F. LEFT INTERNAL JUGULAR VEIN.

G. AZYGOS VEIN.

H. POSTERIOR INTERCOSTAL VEIN.

I. ACCESSORY HEMIAZYGOS VEIN.

J. HEMIAZYGOS VEIN.

K. THORACIC LYMPHATIC DUCT.

L. CISTERNA CHYLI.

M. DIAPHRAGM.

N. INFERIOR VENA CAVA HIATUS.

O. SYMPATHETIC CHAIN.

P. INTERCOSTAL NERVE.

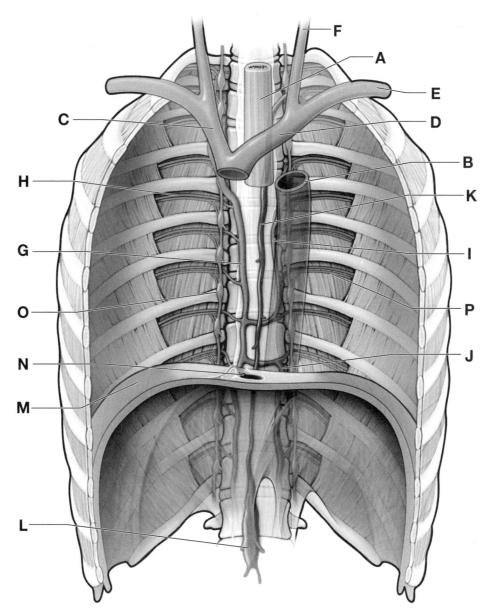

Posterior mediastinum; anterior view.

5d   Superior Mediastinum

The superior mediastinum is a thoroughfare for vessels, nerves, and lymphatics between the neck, upper limbs, and thorax. The trachea has 18 to 20 incomplete hyaline cartilaginous rings, which are open posteriorly and prevent the trachea from collapsing during exhalation. The trachea bifurcates into the right and left bronchial tree at the level of the sternal angle (T4–T5 vertebral level). The ascending aorta arises from the heart at the T4 vertebral level and ascends into the superior mediastinum over the pulmonary vessels and left primary bronchus to become the aortic arch. The aortic arch terminates at the T4 vertebral level to become the thoracic (descending) aorta.

A. RIGHT BRACHIOCEPHALIC VEIN.

B. RIGHT INTERNAL JUGULAR VEIN.

C. RIGHT SUBCLAVIAN VEIN.

D. LEFT BRACHIOCEPHALIC VEIN.

E. AORTA.

F. BRACHIOCEPHALIC TRUNK.

G. LEFT COMMON CAROTID ARTERY.

H. LEFT SUBCLAVIAN ARTERY.

I. LIGAMENTUM ARTERIOSUM.

J. PULMONARY TRUNK.

K. PULMONARY ARTERIES.

L. RIGHT BRONCHIAL TREE.

M. LEFT BRONCHIAL TREE.

N. ESOPHAGUS.

O. TRACHEA.

P. LEFT VAGUS NERVE.

Q. LEFT RECURRENT LARYNGEAL NERVE.

R. RIGHT VAGUS NERVE.

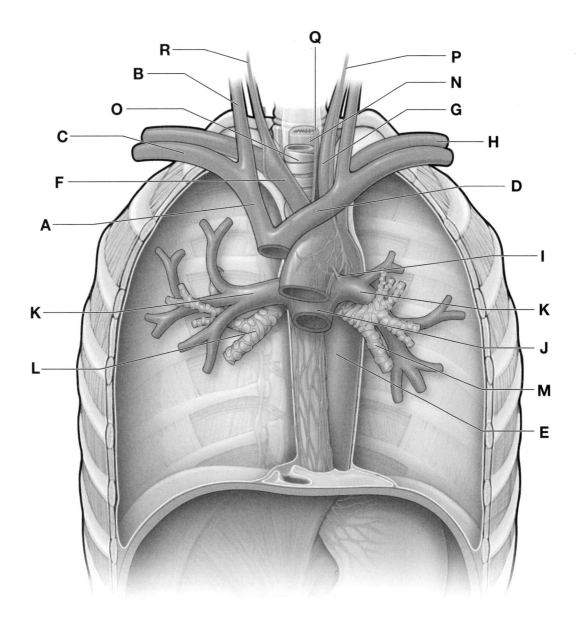

Superior mediastinum; anterior view.

SECTION III

ABDOMEN, PELVIS, AND PERINEUM

6.

Overview of the Abdomen, Pelvis, and Perineum

6a   Osteology of Pelvis

In the adult, the pelvis (os coxae) is formed by the fusion of three bones: ilium, ischium, and pubis. The union of these three bones occurs at the acetabulum. The paired os coxae articulate posteriorly with the sacrum and anteriorly with the pubic symphysis.

The female pelvis differs from the male pelvis because of its importance in childbirth. The pubic arch is the angle between adjacent ischiopubic rami. A typical female pubic arch is usually larger (80 degrees) than the male pubic arch (60 degrees). The angle formed by the female pubic arch can be estimated by the angle between the thumb and the forefinger; in contrast, the male pubic arch is estimated by the angle between the index and middle fingers.

A. ILIUM.

B. PUBIS.

C. ISCHIUM.

D. OBTURATOR FORAMEN.

E. FEMALE PUBIC ARCH (**~80°**).

F. MALE PUBIC ARCH (**~60°**).

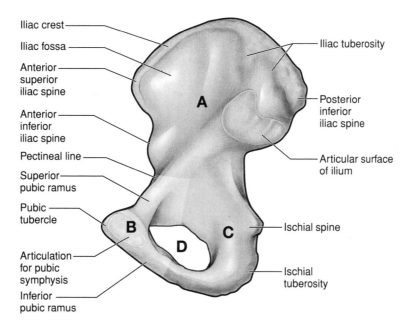

Iliac crest

Iliac fossa

Anterior superior iliac spine

Anterior inferior iliac spine

Pectineal line

Superior pubic ramus

Pubic tubercle

Articulation for pubic symphysis

Inferior pubic ramus

Iliac tuberosity

Posterior inferior iliac spine

Articular surface of ilium

Ischial spine

Ischial tuberosity

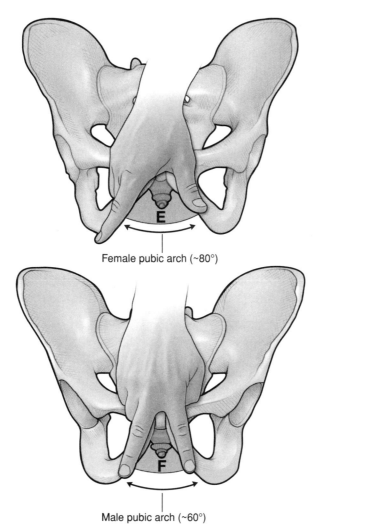

Female pubic arch (~80°)

Male pubic arch (~60°)

Medial view of the os coxa and anterior views of the female and male pelvises.

7.

Anterior Abdominal Wall

7a   Abdomen Surface Anatomy

The abdomen typically is described topographically using the following two methods.

Four Quadrant partitions. The most direct method of partitioning the abdomen is through two intersecting, imaginary planes through the umbilicus that divide the abdomen into four quadrants. The four-quadrant system is straightforward when used to describe anatomic location (i.e., the appendix is located in the lower right quadrant of the abdomen).

A. UPPER RIGHT QUADRANT.

B. UPPER LEFT QUADRANT.

C. LOWER RIGHT QUADRANT.

D. LOWER LEFT QUADRANT.

Nine Regional partitions. For a more precise description, the abdomen is partitioned into nine regions created by two imaginary vertical planes and two imaginary horizontal planes. Paired vertical planes correspond to the midclavicular lines, which descend to the midinguinal point. The upper horizontal plane courses inferior to the costal margin. The lower horizontal plane courses between the two tubercles of the iliac crest. As mentioned before, this system provides a more descriptive location (i.e., gastric ulcers refer pain to the epigastric region).

E. RIGHT HYPOCHONDRIAC REGION.

F. EPIGASTRIC REGION.

G. LEFT HYPOCHONDRIAC REGION.

H. RIGHT LUMBAR REGION.

I. UMBILICAL REGION.

J. LEFT LUMBAR REGION.

K. RIGHT ILIAC REGION.

L. HYPOGASTRIC REGION.

M. LEFT ILIAC REGION.

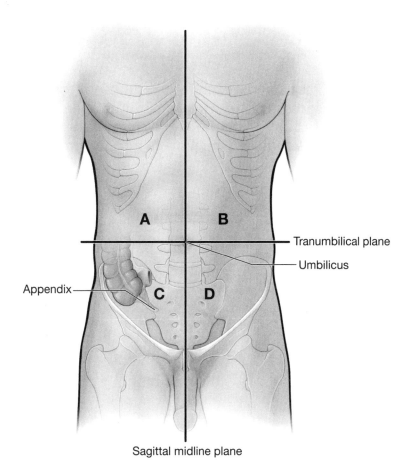

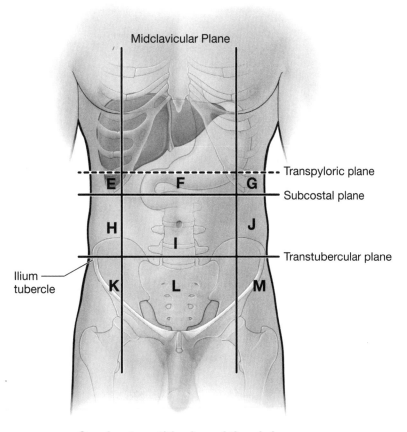

Quadrant partitioning of the abdomen.

7b   Abdominal Wall Layers

Multiple layers of fascia and muscle form the anterior abdominal wall. The layers, from superficial to deep, are skin, two layers of superficial fascia, three layers of muscles and their aponeuroses, transversalis fascia, extraperitoneal fat, and the parietal peritoneum.

A. SKIN.

B. CAMPER FASCIA. Adipose tissue layer of superficial fascia.

C. SCARPA FASCIA. Dense collagenous connective tissue layer of superficial fascia.

D. EXTERNAL OBLIQUE MUSCLE.

E. INTERNAL OBLIQUE MUSCLE.

F. TRANSVERSE ABDOMINIS MUSCLE.

G. TRANSVERSALIS FASCIA.

H. EXTRAPERITONEAL FAT.

I. PARIETAL PERITONEUM.

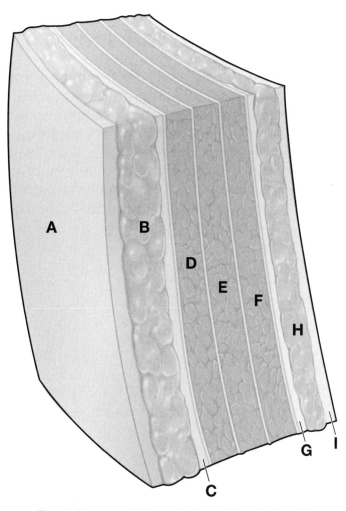

Fascial layers of the anterior abdominal wall.

7c   Anterior Abdominal Wall Muscles

Five paired anterior abdominal wall muscles are deep to the superficial fascia. The external oblique, internal oblique, and transverse abdominis muscles, with their associated aponeuroses, course anterolaterally, whereas the rectus abdominis and tiny pyramidalis muscles course vertically in the anterior midline. Collectively, these muscles compress the abdominal contents, protect vital organs, and flex and rotate the vertebral column. Each muscle receives segmental motor innervation from the lower intercostal and lumbar spinal nerves.

A. EXTERNAL OBLIQUE MUSCLE.

B. INTERNAL OBLIQUE MUSCLE.

C. TRANSVERSE ABDOMINIS MUSCLE.

D. RECTUS ABDOMINIS MUSCLE.

E. RECTUS SHEATH.

F. PYRAMIDALIS MUSCLE.

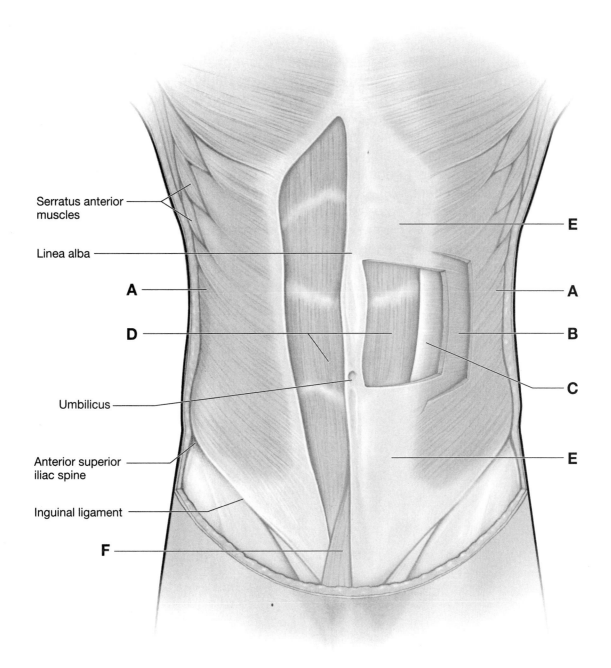

Serratus anterior muscles

Linea alba

A

D

Umbilicus

Anterior superior iliac spine

Inguinal ligament

F

E

A

B

C

E

Step dissection of the anterior abdominal wall muscles.

7d   Anterior Abdominal Wall Muscles (X-Section)

The external oblique, internal oblique, and transverse abdominis aponeuroses envelope the rectus abdominis muscle in a fascial sleeve known as the rectus sheath. The linea alba is a vertical midline of fascia that separates the paired rectus abdominis muscles and is formed by the fusion of the three pairs of anterolateral aponeuroses. The rectus sheath completely encloses the superior three-fourths of the rectus abdominis muscle but only covers the anterior surface of the inferior one-fourth of the muscle. This demarcation region in the rectus sheath is known as the arcuate line. The arcuate line is located midway between the umbilicus and pubic bone and serves also as the site where the inferior epigastric vessels enter to the rectus sheath. Inferior to the arcuate line, the rectus abdominis muscle is in direct contact with the transversalis fascia because the rectus sheath only covers the anterior surface of the rectus abdominis muscle.

A. EXTERNAL OBLIQUE MUSCLE.

B. INTERNAL OBLIQUE MUSCLE.

C. TRANSVERSE ABDOMINIS MUSCLE.

D. RECTUS ABDOMINIS MUSCLE.

E. RECTUS SHEATH.

F. LINEA ALBA.

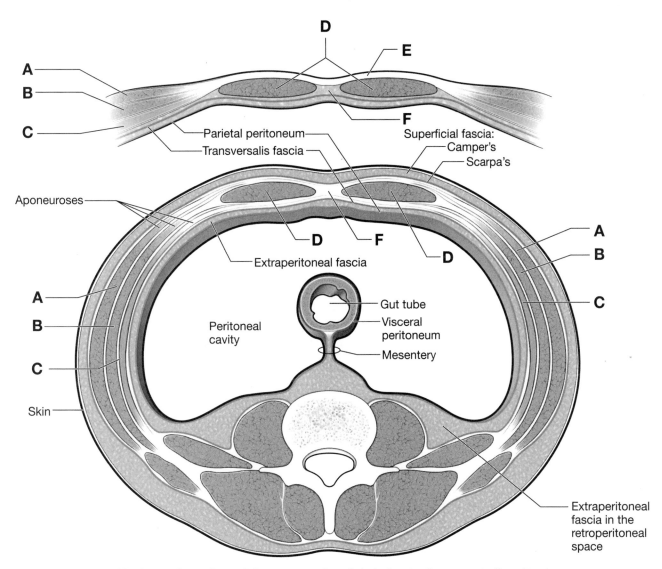

**Horizontal section of the rectus sheath inferior to the arcuate line (top)
and superior to the arcuate line (bottom).**

7e Anterior Abdominal Wall Muscles (X-Section)

The inguinal canal is an oblique passage through the inferior region of the anterior abdominal wall. The inguinal canal is clinically more important in males because it is the passageway for structures to course to and from the testis to the abdomen. The inguinal canal is approximately 5 cm long and extends from the deep inguinal ring downward and medially to the superficial inguinal ring. The inguinal canal lies parallel to and immediately superior to the inguinal ligament. In males, the spermatic cord and scrotum consist of the same layers of muscle and fascia as does the anterior abdominal wall.

During embryonic development, the testes begin development in the region of the kidneys and descend throughout development until they traverse the inguinal canal, protruding through the inferior portion of the anterior abdominal wall. This developmental migration of the testes through the anterior abdominal wall is the basis of formation of the spermatic cord and scrotum from the muscle and fascial layers of the anterior abdominal wall.

A. EXTERNAL OBLIQUE MUSCLE.

B. INTERNAL OBLIQUE MUSCLE.

C. TRANSVERSE ABDOMINIS MUSCLE.

D. RECTUS ABDOMINIS MUSCLE.

E. PYRAMIDALIS MUSCLE.

F. TRANSVERSALIS FASCIA.

G. EXTRAPERITONEAL FASCIA.

H. PARIETAL PERITONEUM.

I. DARTOS FASCIA.

J. EXTERNAL SPERMATIC FASCIA.

K. CREMASTERIC FASCIA AND MUSCLE.

L. INTERNAL SPERMATIC FASCIA.

M. ILIOINGUINAL NERVE.

N. TESTICULAR ARTERY.

O. PAMPINIFORM PLEXUS OF VEINS.

P. DUCTUS DEFERENS.

Q. TESTIS.

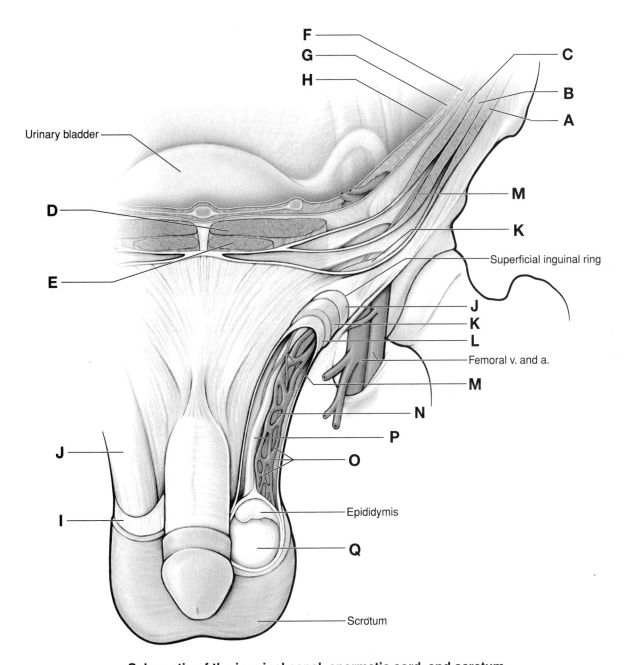

Urinary bladder

Superficial inguinal ring

Femoral v. and a.

Epididymis

Scrotum

Schematic of the inguinal canal, spermatic cord, and scrotum.

8.

Serous Membranes of the Abdominal Cavity

8a	Peritoneum

The abdominopelvic cavity is lined with a serous membrane called the peritoneum. This membrane expands from the internal surface of the abdominal wall to completely or partially surround organs of the abdominopelvic cavities. The peritoneum consists of two layers: parietal peritoneum and visceral peritoneum.

The peritoneal membranes produce a serous fluid that lubricates the peritoneal surfaces, enabling the intraperitoneal organs to slide across one another with minimal friction.

A. PARIETAL PERITONEUM.

B. JEJUNUM AND ILEUM COVERED BY VISCERAL PERITONEUM.

C. GREATER OMENTUM.

D. LESSER OMENTUM.

E. LESSER SAC.

F. GREATER SAC.

G. THE MESENTERY.

H. LIVER.

I. ESOPHAGUS.

J. STOMACH.

K. DUODENUM.

L. PANCREAS.

M. ABDOMINAL AORTA.

N. BLADDER.

O. RECTUM.

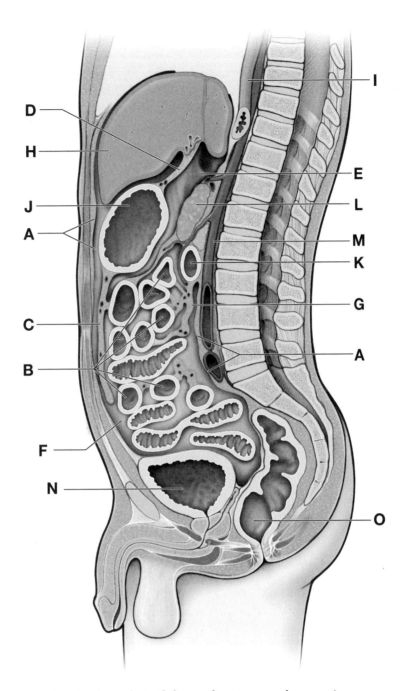

Sagittal section of the peritoneum and mesentery.

8b   Mesentery and Lesser Sac

The parietal peritoneum reflects off of the posterior abdominal wall, forming a fused, double layer of peritoneum surrounding the blood vessels, nerves, and lymphatics to abdominal organs. This double layer of peritoneum, known as the mesentery, suspends the jejunum and ileum from the posterior abdominal wall. The peritoneum that surrounds the gut tube is called the visceral peritoneum.

A. PARIETAL PERITONEUM.

B. VISCERAL PERITONEUM.

C. THE MESENTERY.

D. INTRAPERITONEAL ORGAN.

E. PERITONEAL CAVITY.

F. ARTERY.

G. VEIN.

H. NERVE.

I. LYMPHATICS.

J. GREATER SAC PORTION OF PERITONEAL CAVITY.

K. LESSER SAC PORTION OF PERITONEAL CAVITY.

L. LIVER.

M. STOMACH.

N. SPLEEN.

O. INFERIOR VENA CAVA.

P. ABDOMINAL AORTA.

Q. KIDNEY.

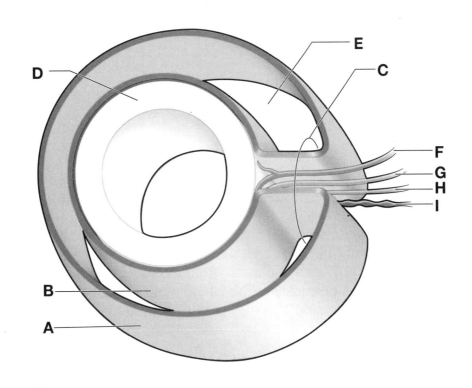

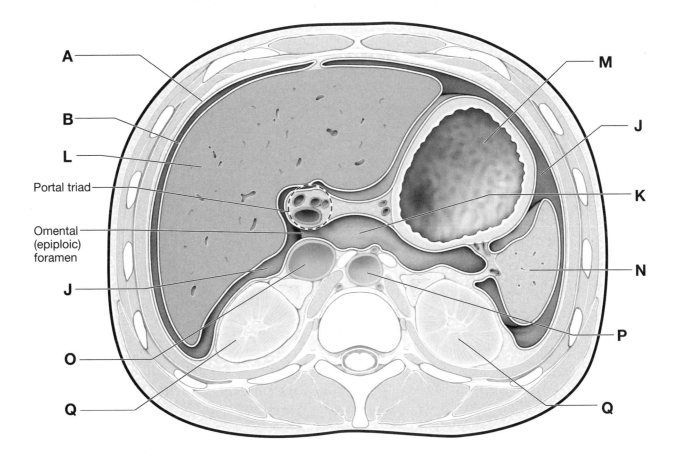

Relationship of the mesentery and neurovascular supply to the intraperitoneal organs and axial section through the liver.

9.

Foregut

9a   Foregut Overview

The foregut consists of the distal end of the esophagus, the stomach, and a portion of the duodenum. In addition, the pancreas, liver, and gallbladder form embryologically from the foregut and thus also are included in this discussion. The celiac trunk is the principal (but not exclusive) artery supplying the foregut. The celiac trunk arises from the abdominal aorta.

A. STOMACH.

B. DUODENUM.

C. LIVER.

D. GALL BLADDER.

E. SPLEEN.

F. GREATER OMENTUM.

G. TRANSVERSE COLON.

H. PANCREAS.

I. INFERIOR VENA CAVA.

J. ABDOMINAL AORTA.

K. CELIAC TRUNK.

L. LEFT GASTRIC ARTERY.

M. COMMON HEPATIC ARTERY.

N. RIGHT GASTRIC ARTERY.

O. GASTRODUODENAL ARTERY.

P. RIGHT GASTRO-OMENTAL ARTERY.

Q. HEPATIC ARTERY PROPER.

R. SPLENIC ARTERY.

S. LEFT GASTRO-OMENTAL ARTERY.

T. COMMON BILE DUCT.

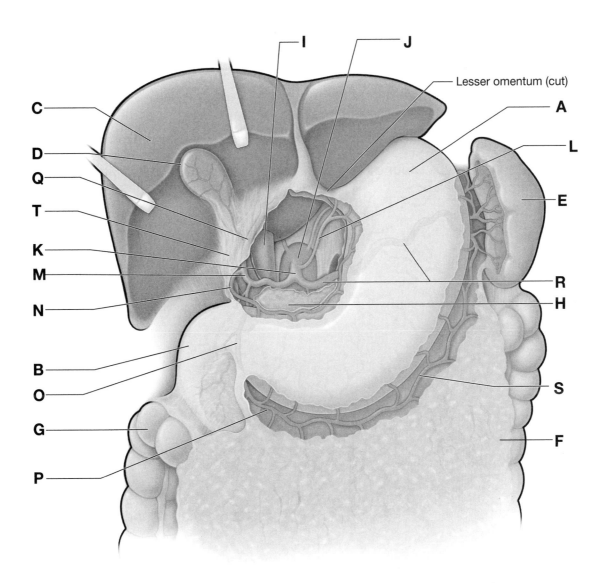

Lesser omentum (cut)

Anterior view of the foregut with the lesser omentum removed.

9b Stomach and Duodenum

The stomach is a portion of the foregut between the esophagus and the duodenum. It is intraperitoneal and located in the upper left quadrant of the abdomen, with the spleen, pancreas, and aorta located deep to the stomach body. The duodenum is approximately 25 cm long and curves around the pancreatic head. All but the first part is retroperitoneal.

A. STOMACH.

The stomach is partitioned into the following regions:

B. CARDIA.

C. FUNDUS.

D. BODY.

E. PYLORUS.

F. DUODENUM.

The duodenum is divided into the following four parts:

G. PART ONE (SUPERIOR).

H. PART TWO (DESCENDING). Possesses the major duodenal papilla.

I. PART THREE (HORIZONTAL). Crossed by SMA and SMV.

J. PART FOUR (ASCENDING).

K. ESOPHAGUS.

L. SUPERIOR MESENTERIC VEIN (SMV).

M. SUPERIOR MESENTERIC ARTERY (SMA).

N. GALL BLADDER.

O. CYSTIC DUCT.

P. HEPATIC DUCTS.

Q. COMMON BILE DUCT.

R. PANCREAS.

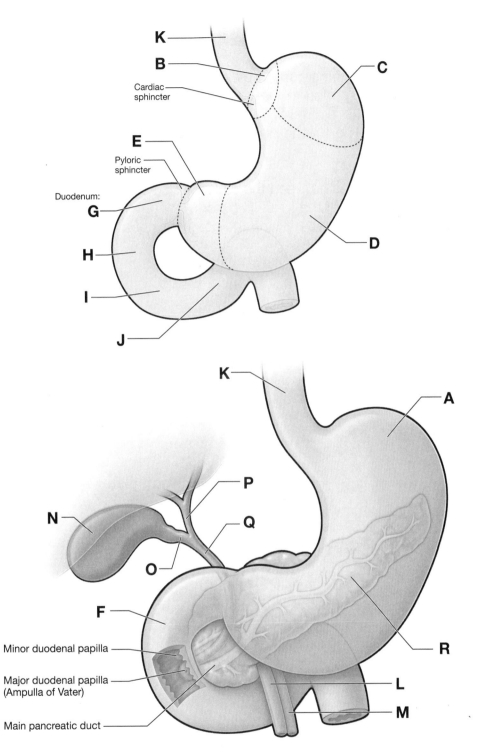

Cardiac sphincter

Pyloric sphincter

Duodenum:

Minor duodenal papilla

Major duodenal papilla (Ampulla of Vater)

Main pancreatic duct

Parts of the stomach and duodenum.

9c   Liver and Portal Triad

In addition to its numerous metabolic activities, the liver secretes bile. Bile is transported to the gallbladder, where it is stored. When food reaches the duodenum, the gallbladder releases bile, which emulsifies fat. The liver is divided into the right, left, quadrate and caudate lobes.

The portal triad lies between the caudate and quadrate lobes and is the structural unit of the liver. The portal triad consists of the portal vein, proper hepatic artery, and the common hepatic duct.

A. RIGHT LOBE OF LIVER.

B. LEFT LOBE OF LIVER.

C. QUADRATE LOBE OF LIVER.

D. CAUDATE LOBE OF LIVER.

E. GALL BLADDER.

F. CYSTIC DUCT.

G. RIGHT HEPATIC DUCT.

H. LEFT HEPATIC DUCT.

I. COMMON HEPATIC DUCT.

J. COMMON BILE DUCT.

K. HEPATIC ARTERY PROPER.

L. PORTAL VEIN.

M. DUODENUM.

N. PANCREAS.

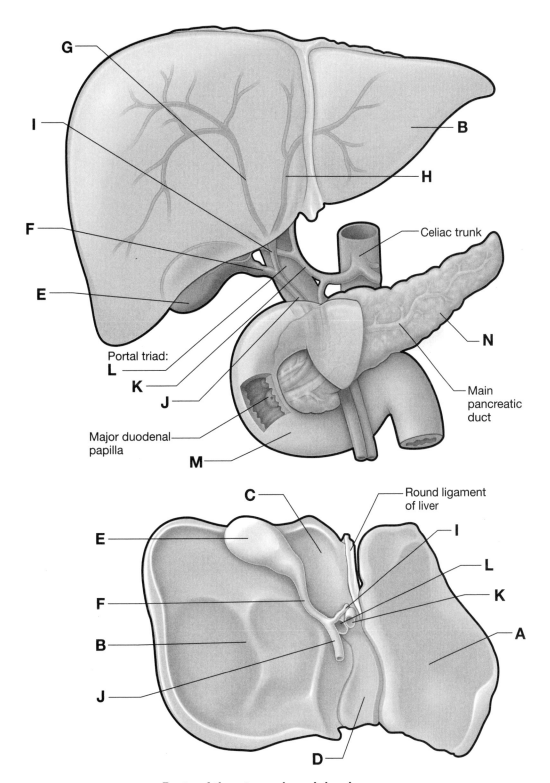

Celiac trunk

B

H

N

Main pancreatic duct

Portal triad:

L

K

J

Major duodenal papilla

M

C

Round ligament of liver

I

L

K

A

D

Parts of the stomach and duodenum.

| 9d | **Foregut—Stomach Removed** |

A. CELIAC TRUNK.

B. LEFT GASTRIC ARTERY.

C. SPLENIC ARTERY.

D. SHORT GASTRIC ARTERY.

E. LEFT GASTRO-OMENTAL ARTERY.

F. COMMON HEPATIC ARTERY.

G. HEPATIC ARTERY PROPER.

H. CYSTIC ARTERY.

I. GASTRODUODENAL ARTERY.

J. SUPERIOR PANCREATICODUODENAL ARTERIES.

K. SUPERIOR MESENTERIC VEIN (SMV).

L. SUPERIOR MESENTERIC ARTERY (SMA).

M. INFERIOR PANCREATICODUODENAL ARTERIES.

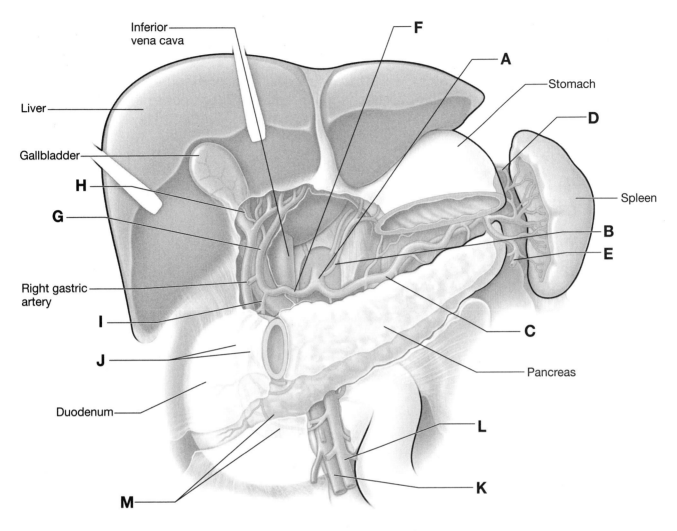

Inferior vena cava

Liver

Gallbladder

Right gastric artery

Duodenum

Stomach

Spleen

Pancreas

F
A
D
B
E
C
H
G
I
J
M
L
K

Parts of the stomach and duodenum.

| 9e | **Pancreas** |

The pancreas is an organ of dual function: exocrine secretion for digestion and endocrine function for the regulation of glucose metabolism. It is a retroperitoneal organ located at the L2 vertebral level.

A. PANCREAS.

B. MAIN PANCREATIC DUCT.

C. DUODENUM.

D. GALL BLADDER.

E. SPLEEN.

F. CELIAC TRUNK.

G. SPLENIC ARTERY.

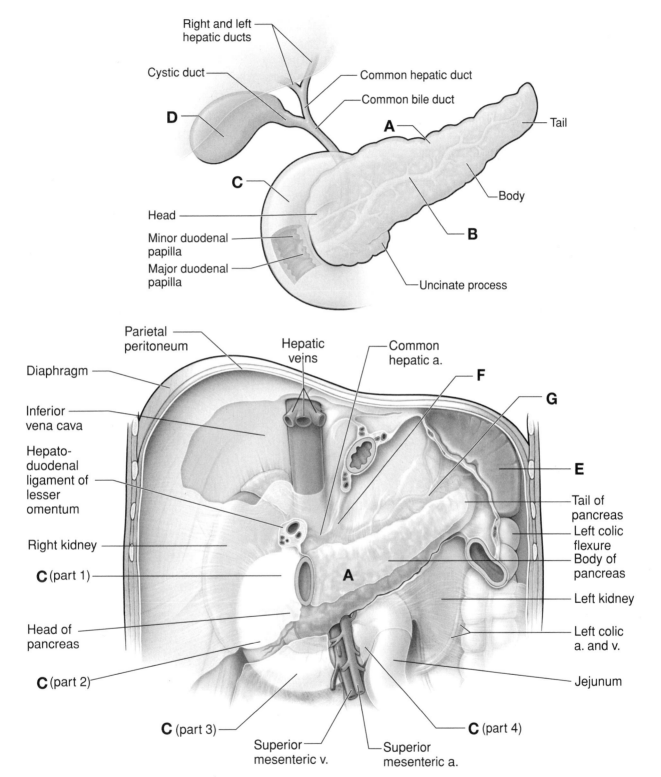

Pancreas.

9f   Celiac Trunk

The principal blood supply to the organs of the foregut is the celiac trunk. The celiac trunk is an unpaired artery arising from the abdominal aorta, immediately below the aortic hiatus of the diaphragm at the T12 vertebral level. The celiac trunk divides into the left gastric, splenic, and common hepatic arteries. The pancreaticoduodenal arteries form an anastomosis between the foregut (superior pancreaticoduodenal artery) and the midgut (inferior pancreaticoduodenal artery).

A. ABDOMINAL AORTA.

B. CELIAC TRUNK.

C. SPLENIC ARTERY.

D. LEFT GASTRO-OMENTAL ARTERY.

E. LEFT GASTRIC ARTERY.

F. COMMON HEPATIC ARTERY.

G. RIGHT GASTRIC ARTERY.

H. HEPATIC ARTERY PROPER.

I. GASTRODUODENAL ARTERY.

J. RIGHT GASTRO-OMENTAL ARTERY.

K. SUPERIOR PANCREATICODUODENAL ARTERY.

L. SUPERIOR MESENTERIC ARTERY (SMA).

M. INFERIOR PANCREATICODUODENAL ARTERY.

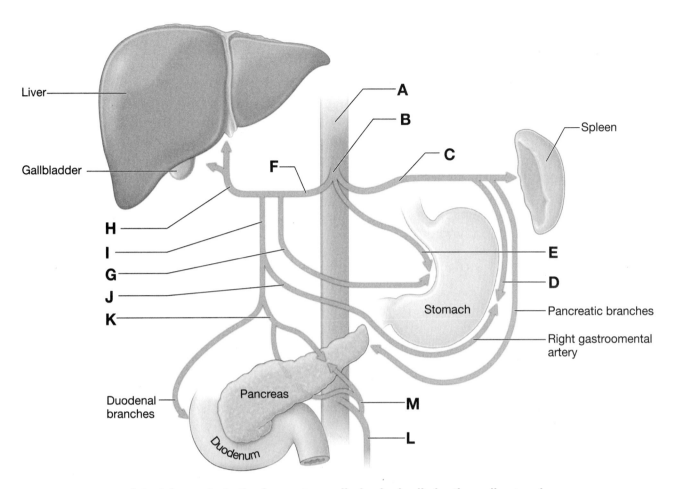

Arterial supply to the foregut supplied principally by the celiac trunk.

9g   Foregut Venous Drainage

The principal venous drainage of the foregut is to the portal venous system. The portal venous system drains nutrient-rich venous blood from the gastrointestinal tract and the spleen to the liver. Three collecting veins converge to form the portal vein: the splenic vein and the superior mesenteric and inferior mesenteric veins. Portal venous blood flows to the liver, where nutrients are metabolized. The metabolic products are collected in central veins, which are tributaries of the hepatic veins. The hepatic veins emerge from the liver to drain into the inferior vena cava.

A. PORTAL VEIN.

B. LEFT GASTRIC VEIN.

C. RIGHT GASTRIC VEIN.

D. SPLENIC VEIN.

E. LEFT GASTRO-OMENTAL VEIN.

F. RIGHT GASTRO-OMENTAL VEIN.

G. SUPERIOR PANCREATICODUODENAL VEIN.

H. INFERIOR PANCREATICODUODENAL VEIN.

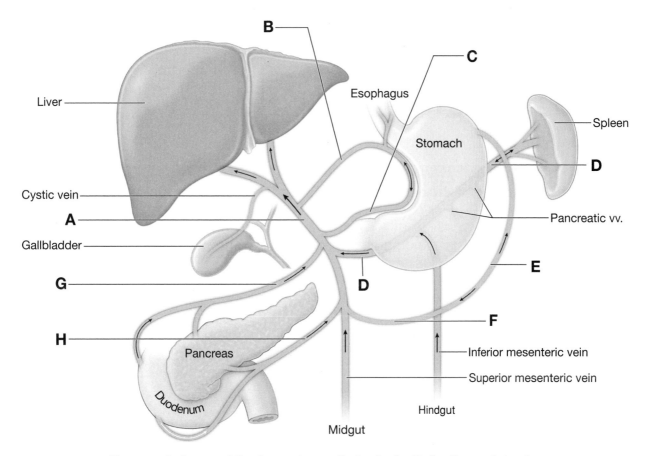

Venous drainage of the foregut supplied principally by the portal vein.

10.

Midgut and Hindgut

10a   Midgut Overview

The midgut consists of the distal half of the duodenum, jejunum, ileum, cecum, ascending colon, and the proximal half of the transverse colon.

A. JEJUNUM.

B. ILEUM.

C. CECUM.

D. ASCENDING COLON.

E. TRANSVERSE COLON.

F. GREATER OMENTUM.

G. MEDIAN UMBILICAL FOLD.

H. MEDIAL UMBILICAL FOLD.

I. LATERAL UMBILICAL FOLD.

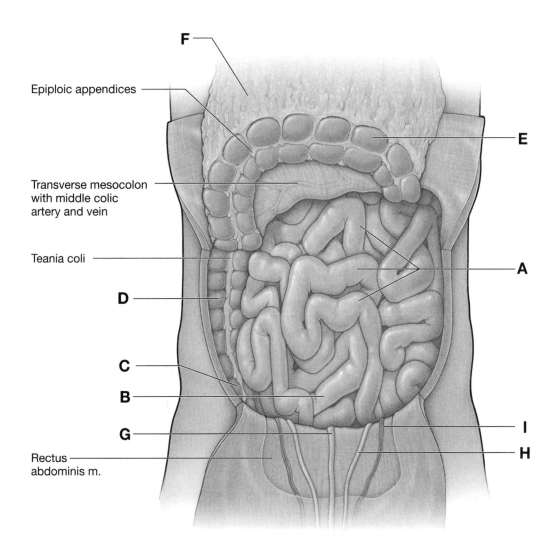

F

Epiploic appendices

E

Transverse mesocolon
with middle colic
artery and vein

Teania coli

A

D

C

B

I

G

H

Rectus
abdominis m.

**Midgut with the greater omentum reflected superiorly and the anterior
abdominal wall reflected inferiorly.**

10b   Midgut Arteries

The midgut receives its arterial supply primarily from branches of the superior mesenteric artery.

A. SUPERIOR MESENTERIC ARTERY.

B. INFERIOR PANCREATICODUODENAL ARTERY.

C. JEJUNAL AND ILEAL ARTERIES.

D. ILEOCOLIC ARTERY.

E. APPENDICULAR ARTERY.

F. RIGHT COLIC ARTERY.

G. MIDDLE COLIC ARTERY.

H. MARGINAL ARTERY (OF DRUMMOND).

I. JEJUNUM.

J. ILEUM.

K. APPENDIX.

L. CECUM.

M. ASCENDING COLON.

N. TRANSVERSE COLON.

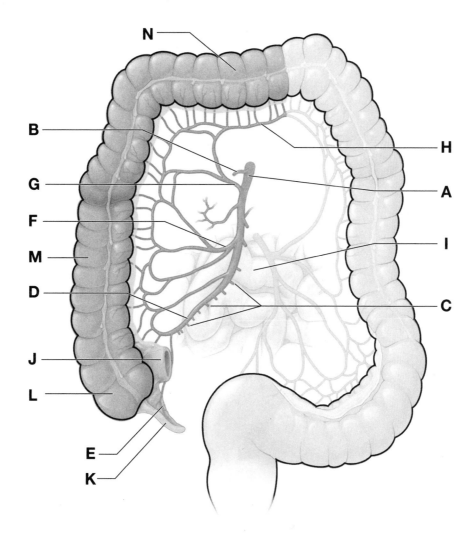

Primary blood supply to the midgut is through the superior mesenteric artery.

10c   Hindgut Overview

The hindgut consists of the distal half of the transverse colon, descending colon, sigmoid colon, and the proximal third of the rectum.

A. TRANSVERSE COLON.

B. DESCENDING COLON.

C. SIGMOID COLON.

D. RECTUM.

E. TRANSVERSE MESOCOLON.

F. SIGMOID MESOCOLON.

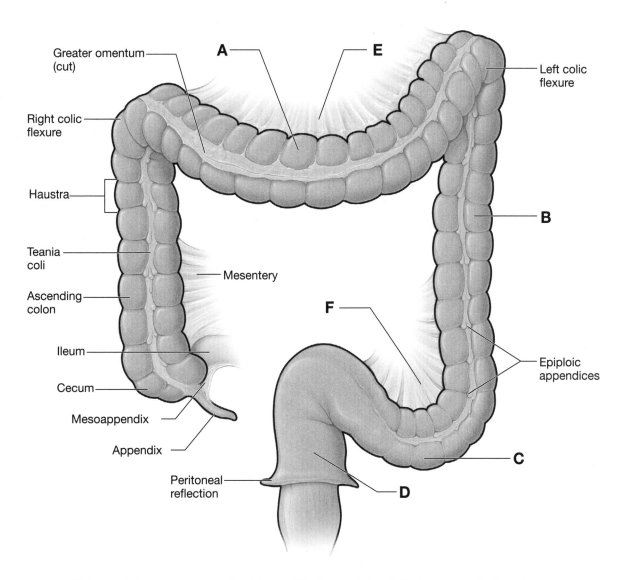

Greater omentum (cut)

Right colic flexure

Haustra

Teania coli

Ascending colon

Ileum

Cecum

Mesoappendix

Appendix

Peritoneal reflection

Left colic flexure

Mesentery

Epiploic appendices

A

E

B

F

C

D

Primary blood supply to the hindgut is through the inferior mesenteric artery.

| **10d** | **Hindgut Arteries** |

Branches of the inferior mesenteric artery and vein provide vascular supply to the hindgut.

A. INFERIOR MESENTERIC ARTERY.

B. LEFT COLIC ARTERY.

C. SIGMOIDAL ARTERIES.

D. SUPERIOR RECTAL ARTERY.

E. MARGINAL ARTERY (OF DRUMMOND).

F. TRANSVERSE COLON.

G. DESCENDING COLON.

H. SIGMOID COLON.

I. RECTUM.

J. ANUS.

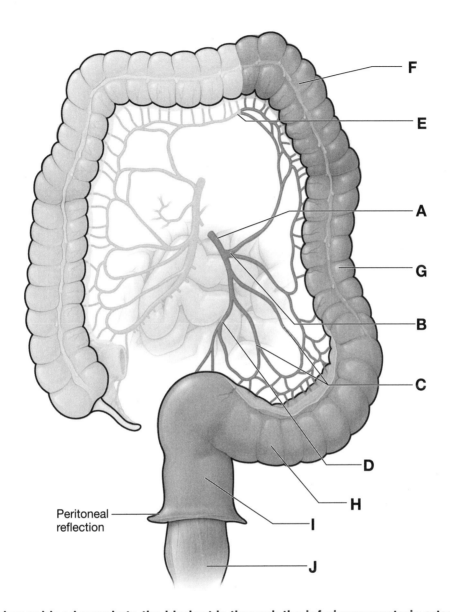

F

E

A

G

B

C

D

H

Peritoneal
reflection

I

J

Primary blood supply to the hindgut is through the inferior mesenteric artery.

10e	**Portal System**

The portal system is responsible for transporting blood from most of the gastrointestinal tract to the liver for metabolic processing before the blood returns to the heart. The portal system drains venous blood from the distal end of the esophagus, stomach, small and large intestines, proximal portion of the rectum, pancreas, and spleen. The portal system is the venous counterpart to areas supplied by the celiac trunk and the superior and inferior mesenteric arteries.

A. PORTAL VEIN.

B. RIGHT GASTRIC VEIN.

C. LEFT GASTRIC VEIN.

D. GASTRO-OMENTAL VEIN.

E. SUPERIOR MESENTERIC VEIN.

F. SPLENIC VEIN.

G. INFERIOR MESENTERIC VEIN.

H. SUPERIOR RECTAL VEIN.

I. HEPATIC VEINS.

J. INFERIOR VENA CAVA.

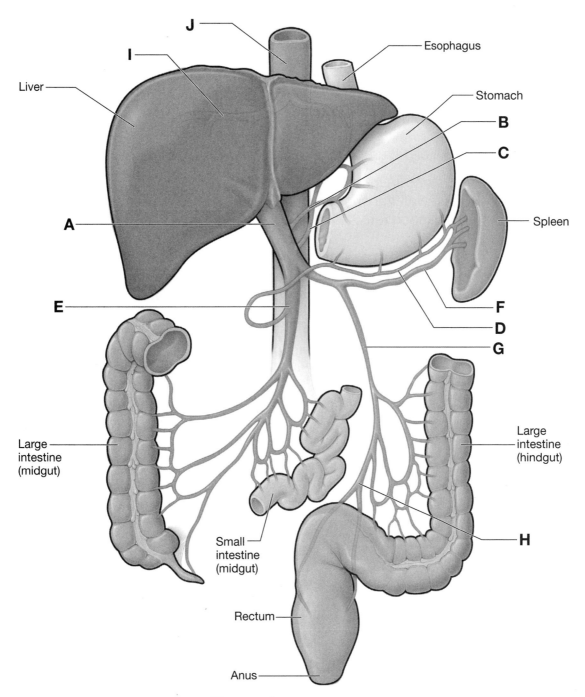

The portal venous system.

10f   Portal–Caval Anastomoses

All veins in the abdomen return blood to the heart via two routes:

- Portal system. Veins from the foregut, midgut, and hindgut drain blood to the liver before the blood enters the inferior vena cava and ultimately returns to the heart.

- Caval system. Veins from the lower limbs, pelvis, and posterior abdominal wall transport blood directly to the inferior vena cava before the blood returns to the heart.

Portal–caval anastomoses are regions of the gastrointestinal tract that are drained by both the portal and systemic (–caval) systems. The principal portal–caval anastomoses are as follows:

1. Distal portion of the esophagus. The left gastric vein and azygos veins drain blood from the distal portion of the esophagus.

2. Anterior abdominal wall. The paraumbilical veins and inferior epigastric veins drain the tissue surrounding the umbilicus.

3. Rectum. The proximal portion of the rectum is drained via the superior rectal vein as well as the middle and inferior rectal veins.

    A. PORTAL VEIN.

    B. AZYGOS VEIN.

    C. ESOPHAGEAL VEIN.

    D. LEFT GASTRIC VEIN.

    E. PARAUMBILICAL VEIN.

    F. INFERIOR EPIGASTRIC VEIN.

    G. EXTERNAL ILIAC VEIN.

    H. INTERNAL ILIAC VEIN.

    I. COMMON ILIAC VEIN.

    J. INFERIOR VENA CAVA.

    K. SUPERIOR RECTAL VEIN.

    L. INFERIOR MESENTERIC VEIN.

    M. SPLENIC VEIN.

    N. MIDDLE AND INFERIOR ILIAC VEINS.

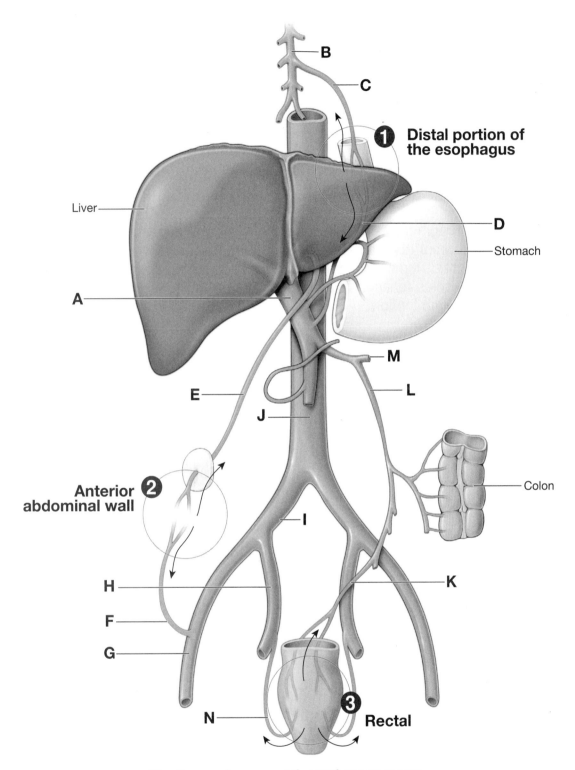

1 Distal portion of the esophagus

2 Anterior abdominal wall

3 Rectal

Liver

Stomach

Colon

The three primary portal–caval anastomoses.

11.

Posterior Abdominal Wall

| 11a | **Posterior Abdominal Wall Muscles** |

The diaphragm forms the superior and much of the posterior border of the posterior abdominal wall. In addition, the psoas major, iliacus, and quadratus lumborum muscles form the posterior abdominal wall. These muscles function in respiration (diaphragm) as well as trunk and lower limb motion.

A. DIAPHRAGM.

B. INFERIOR VENA CAVA.

C. ESOPHAGUS.

D. ABDOMINAL AORTA.

E. TRANSVERSUS ABDOMINIS MUSCLE.

F. QUADRATUS LUMBORUM MUSCLE.

G. ILIACUS MUSCLE.

H. PSOAS MAJOR MUSCLE.

I. ILIOPSOAS MUSCLE.

J. PSOAS MINOR MUSCLE.

K. INGUINAL LIGAMENT.

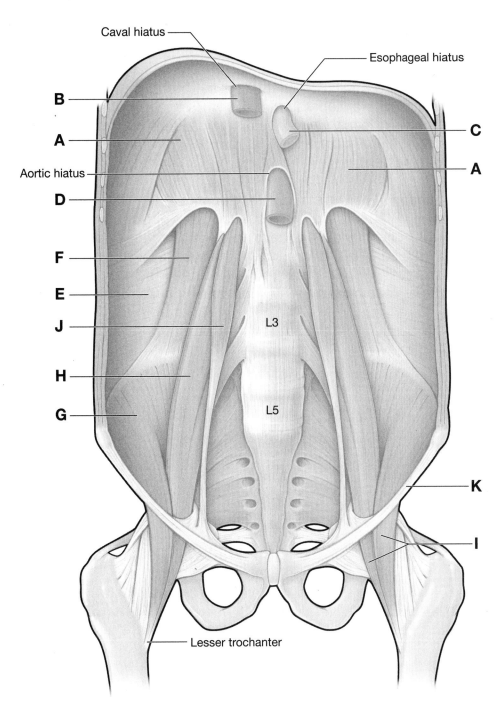

Caval hiatus

Esophageal hiatus

B

A

Aortic hiatus

D

C

A

F

E

J

L3

H

L5

G

K

I

Lesser trochanter

Muscles of the posterior abdominal wall.

11b   Vessels of the Posterior Abdominal Wall

The aorta enters the abdomen from the thorax by traversing the aortic hiatus of the diaphragm at the T12 vertebral level. The aorta courses along the midline, on the anterior surface of vertebral bodies to the left of the inferior vena cava. It supplies the abdominal viscera and body wall.

The inferior vena cava is located to the right of the abdominal aorta. The union of the left and right common iliac veins forms the inferior vena cava. The inferior vena cava ascends along the right side of the vertebral bodies. Before entering the thoracic cavity, the inferior vena cava courses within a groove on the posterior surface of the liver. The inferior vena cava receives tributaries from the abdominal wall.

A. INFERIOR VENA CAVA.

B. HEPATIC VEINS.

C. RENAL VEIN.

D. SUPRARENAL VEIN.

E. COMMON ILIAC VEIN.

F. EXTERNAL ILIAC VEIN.

G. INTERNAL ILIAC VEIN.

H. AORTA.

I. CELIAC TRUNK.

J. SUPERIOR MESENTERIC ARTERY.

K. INFERIOR MESENTERIC ARTERY.

L. COMMON ILIAC ARTERY.

M. EXTERNAL ILIAC ARTERY.

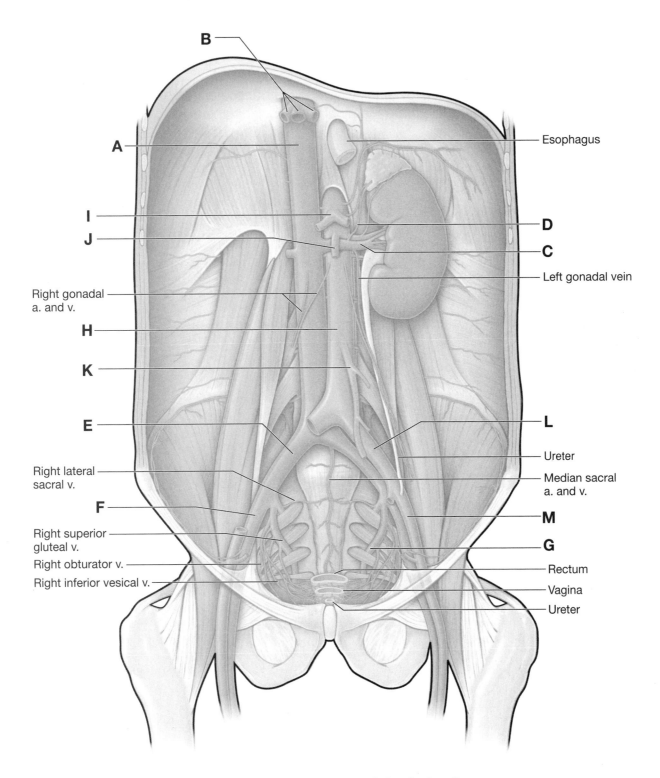

B

A ———————————————— Esophagus

I

J

Right gonadal
a. and v.

H

K

D

C

Left gonadal vein

E

Right lateral
sacral v.

F

L

Ureter

Median sacral
a. and v.

M

Right superior
gluteal v.

Right obturator v.

Right inferior vesical v.

G

Rectum

Vagina

Ureter

Vessels of the posterior abdominal wall.

11c   Nerves of the Posterior Abdominal Wall

The ventral rami of the lower thoracic and lumbar spinal nerves provide somatic innervation to the abdominal wall muscles and skin.

A. SYMPATHETIC TRUNK.

B. SUBCOSTAL NERVE.

C. ILIOHYPOGASTRIC NERVE.

D. ILIOINGUINAL NERVE.

E. GENITOFEMORAL NERVE.

F. LATERAL FEMORAL CUTANEOUS NERVE.

G. FEMORAL NERVE.

H. OBTURATOR NERVE.

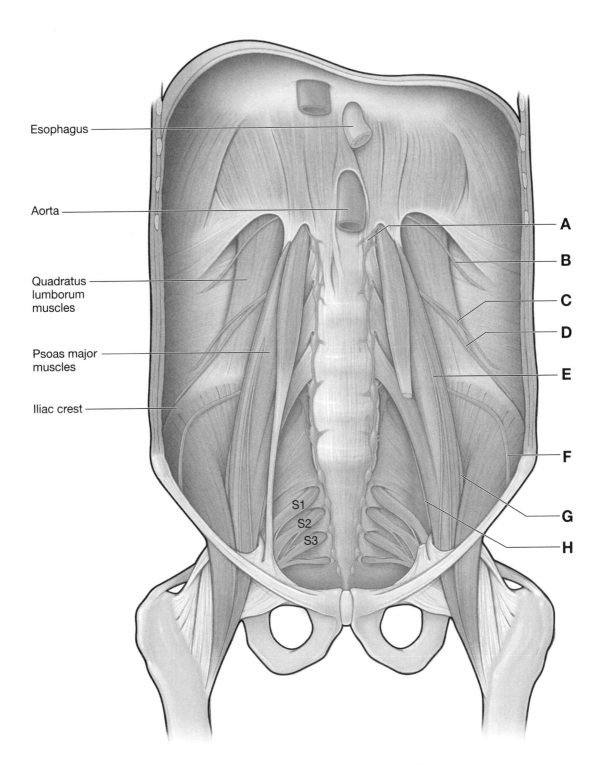

Esophagus

Aorta

Quadratus
lumborum
muscles

Psoas major
muscles

Iliac crest

S1
S2
S3

A
B
C
D
E
F
G
H

Nerves of the posterior abdominal wall.

11d   Autonomics of the Posterior Abdominal Wall

The prevertebral plexus is a network of sympathetic and parasympathetic fibers that innervate the digestive, urinary, and reproductive systems. Sympathetic nerves contribute to the prevertebral plexus via splanchnic nerves from the sympathetic trunk. Parasympathetic nerves contribute to the prevertebral plexus via the vagus nerve [cranial nerve (CN X)] and pelvic splanchnics from the S2–S4 spinal nerves.

A. GREATER SPLANCHNIC NERVE.

B. LESSER SPLANCHNIC NERVE.

C. CELIAC GANGLION.

D. SUPERIOR MESENTERIC GANGLION.

E. AORTICORENAL GANGLIA.

F. INFERIOR MESENTERIC GANGLION.

G. SYMPATHETIC TRUNK.

H. VAGUS NERVE.

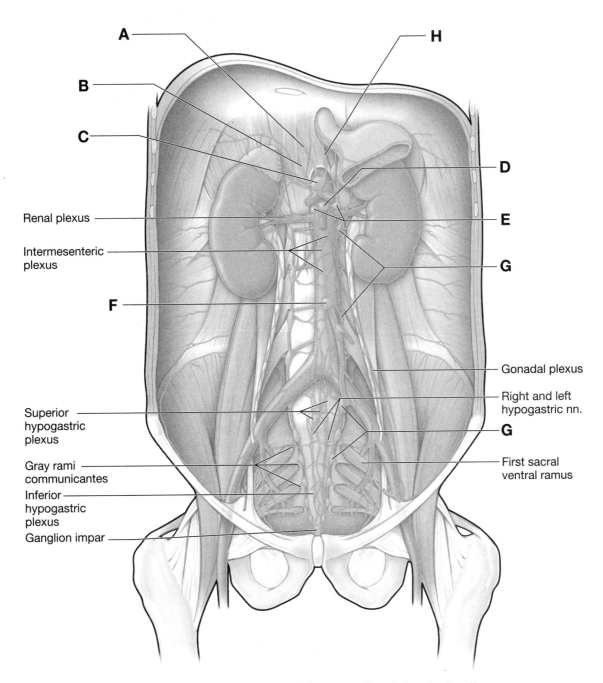

A

H

B

C

D

E

Renal plexus

Intermesenteric
plexus

G

F

Gonadal plexus

Right and left
hypogastric nn.

Superior
hypogastric
plexus

G

Gray rami
communicantes

First sacral
ventral ramus

Inferior
hypogastric
plexus

Ganglion impar

Autonomic nerves of the posterior abdominal wall.

11e   **Autonomics of the Posterior Abdominal Wall**

The prevertebral (preaortic) plexus is a network of autonomic nerve fibers covering the abdominal aorta and extending into the pelvic cavity between the common iliac arteries. This plexus serves as a common pathway for the following autonomics:

A. ANTERIOR VAGAL TRUNK.

B. POSTERIOR VAGAL TRUNK.

C. PELVIC SPLANCHNIC NERVES.

D. GREATER, LESSER, AND LEAST SPLANCHNIC NERVES.

E. CELIAC GANGLIA AND PLEXUS.

F. SUPERIOR MESENTERIC GANGLION AND PLEXUS.

G. INFERIOR MESENTERIC GANGLION AND PLEXUS.

H. SYMPATHETIC TRUNK AND GANGLION.

I. LUMBAR SPLANCHNIC NERVE.

J. RAMI COMMUNICANTES.

K. SACRAL SPLANCHNIC NERVES.

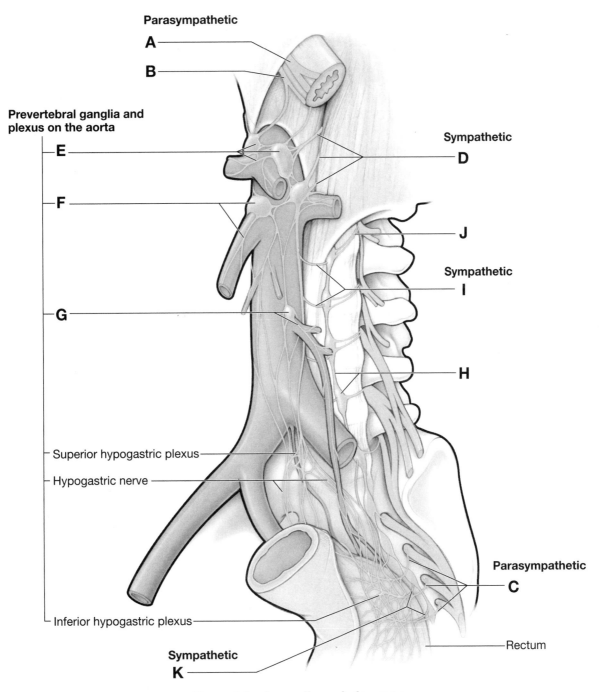

Parasympathetic
A
B

**Prevertebral ganglia and
plexus on the aorta**
E
F
G

Superior hypogastric plexus
Hypogastric nerve

Inferior hypogastric plexus

Sympathetic
K

Sympathetic
D

J

Sympathetic
I

H

Parasympathetic
C

Rectum

Prevertebral ganglia and plexuses.

11f   Adrenal Glands, Kidneys, and Ureters

The adrenal (suprarenal) glands are responsible for regulating stress through the production and secretion of hormones such as adrenalin (epinephrine), glucocorticoids, mineralocorticoids, and androgens. The kidneys filter systemic blood to produce urine, which is transported to the bladder by the ureters.

A. ADRENAL GLAND.

B. CORTEX.

C. MEDULLA.

D. KIDNEY.

E. RENAL VEIN.

F. RENAL ARTERY.

G. URETER.

H. BLADDER.

I. PERINEAL FAT.

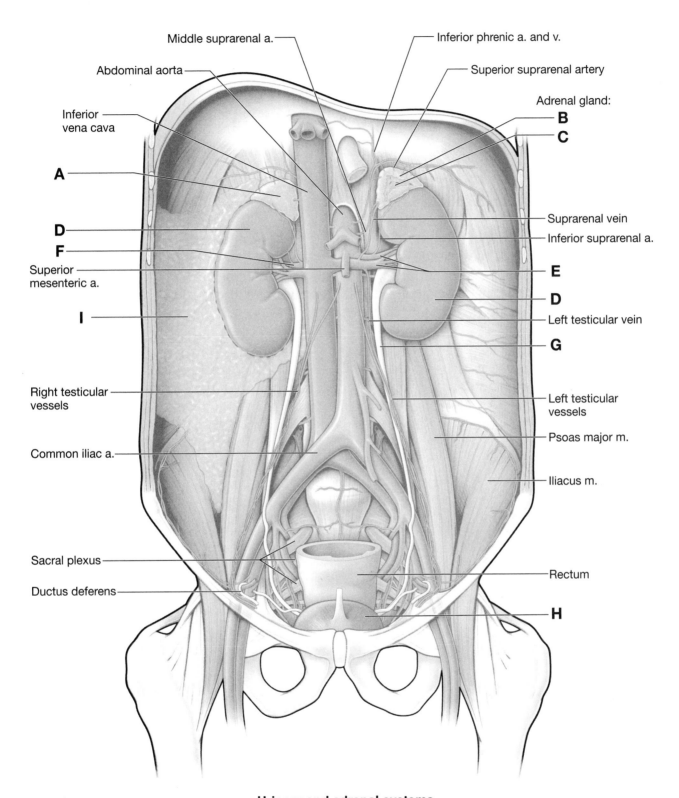

Middle suprarenal a.

Inferior phrenic a. and v.

Abdominal aorta

Superior suprarenal artery

Inferior vena cava

Adrenal gland:

B

C

A

Suprarenal vein

D

Inferior suprarenal a.

F

Superior mesenteric a.

E

D

I

Left testicular vein

G

Right testicular vessels

Left testicular vessels

Psoas major m.

Common iliac a.

Iliacus m.

Sacral plexus

Rectum

Ductus deferens

H

Urinary and adrenal systems.

12.

Pelvis and Perineum

12a   Pelvic Floor Muscles

The pelvic diaphragm forms the floor of the pelvis and is formed by the union of the levator ani and the coccygeus muscles. A layer of fascia lines the superior and inferior aspects of the pelvic diaphragm. The levator ani muscle consists of three separate muscles: pubococcygeus, puborectalis, and iliococcygeus.

The pelvic diaphragm circumferentially attaches along the pubis, lateral pelvic walls, and coccyx. The rectum pierces the center of the pelvic diaphragm, giving the appearance of a funnel suspended within the pelvis. In addition to the rectum, the urethra and the vagina (in females) and the urethra (in males) pierce the pelvic diaphragm.

A. COCCYGEUS MUSCLE.

B. ILIOCOCCYGEUS MUSCLE.

C. PUBOCOCCYGEUS MUSCLE.

D. PUBORECTALIS MUSCLE.

E. OBTURATOR INTERNUS MUSCLE.

F. PIRIFORMIS MUSCLE.

G. ISCHIOCAVERNOSUS MUSCLE.

H. SUPERFICIAL TRANSVERSE PERINEAL MUSCLE.

I. UROGENITAL DIAPHRAGM.

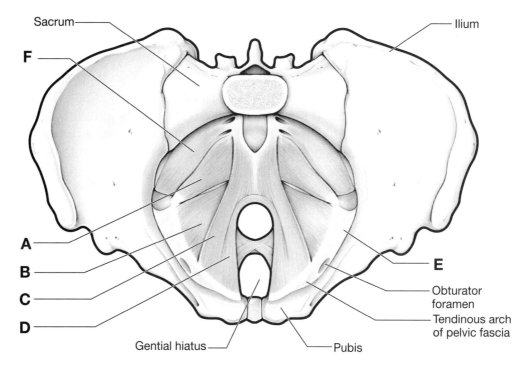

Sacrum

Ilium

F

A

B

C

D

E

Obturator foramen

Tendinous arch of pelvic fascia

Gential hiatus

Pubis

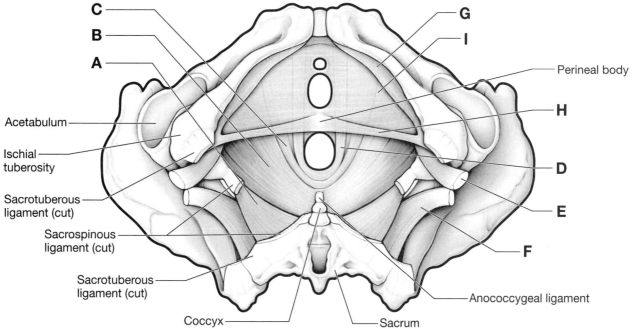

C

B

A

G

I

Perineal body

Acetabulum

Ischial tuberosity

Sacrotuberous ligament (cut)

Sacrospinous ligament (cut)

Sacrotuberous ligament (cut)

H

D

E

F

Anococcygeal ligament

Coccyx

Sacrum

Muscles of the female pelvic floor. Superior view (top) and inferior view (bottom).

12b   Male Perineum

The perineum is the diamond-shaped region inferior to the pelvic diaphragm. The pubic symphysis, pubic arches, ischial tuberosities, and coccyx bound the perineum. An imaginary line between the ischial tuberosities divides the perineum into an anterior (urogenital) triangle and a posterior (ischioanal) triangle. The urogenital triangle extends between the paired ischiopubic rami. The ischioanal triangle is the fat-filled area surrounding the anal canal.

A. BULBOSPONGIOSUS MUSCLE.

B. ISCHIOCAVERNOSUS MUSCLE.

C. SUPERFICIAL TRANSVERSE PERINEAL MUSCLE.

D. LEVATOR ANI MUSCLE.

E. OBTURATOR INTERNUS MUSCLE.

F. SUPERFICIAL PERINEAL FASCIA.

G. OBTURATOR FASCIA.

H. DEEP TRANSVERSE PERINEAL MUSCLE.

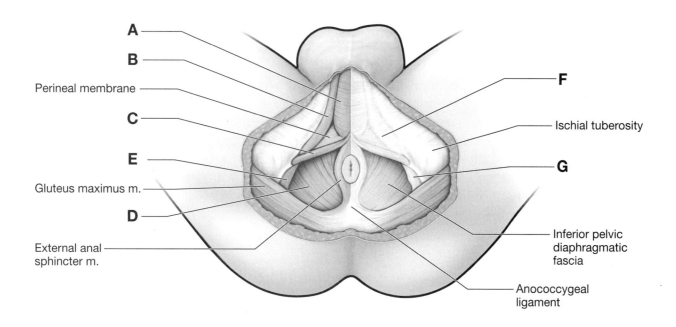

A

B

Perineal membrane

C

E

Gluteus maximus m.

D

External anal
sphincter m.

F

Ischial tuberosity

G

Inferior pelvic
diaphragmatic
fascia

Anococcygeal
ligament

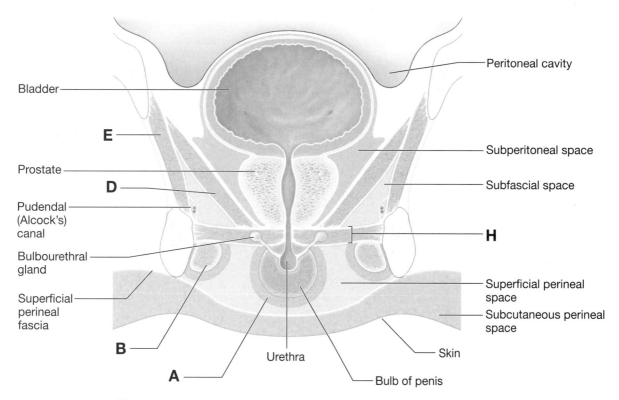

Bladder

E

Prostate

D

Pudendal
(Alcock's)
canal

Bulbourethral
gland

Superficial
perineal
fascia

B

A

Urethra

Peritoneal cavity

Subperitoneal space

Subfascial space

H

Superficial perineal
space

Subcutaneous perineal
space

Skin

Bulb of penis

**Male perineum. Perineal view (upper) and coronal section
through pelvis and perineum (lower).**

12c   Pelvic Vasculature

The common iliac arteries divide at the sacroiliac joints and become the external and internal iliac arteries. The external iliac arteries mainly serve the lower limb. The internal iliac arteries distribute blood to the pelvic walls and viscera (i.e., rectum, bladder, prostate, ductus deferens, uterus, and uterine tubes). In addition, these arteries distribute blood to the gluteal region, the perineum, and the medial compartment of the thigh.

A. COMMON ILIAC ARTERY.

B. EXTERNAL ILIAC ARTERY.

C. INTERNAL ILIAC ARTERY.

D. ILIOLUMBAR ARTERY.

E. SUPERIOR GLUTEAL ARTERY.

F. LATERAL SACRAL ARTERY.

G. OBLITERATED UMBILICAL ARTERY.

H. OBTURATOR ARTERY.

I. UTERINE ARTERY.

J. INFERIOR GLUTEAL ARTERY.

K. MIDDLE RECTAL ARTERY.

L. INTERNAL PUDENDAL ARTERY.

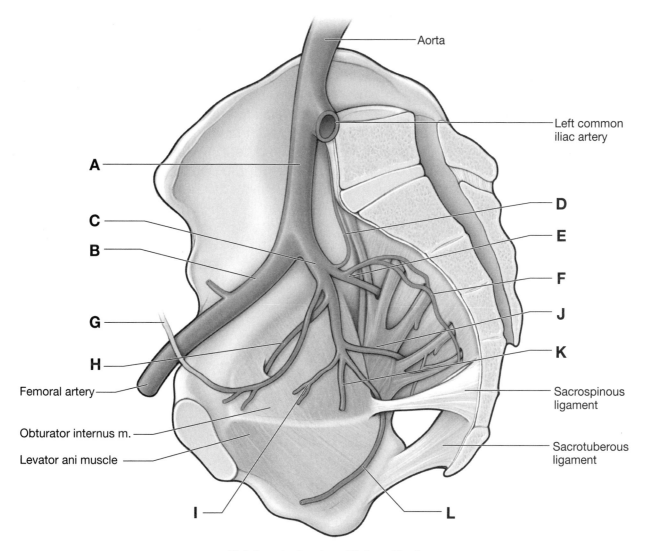

Pelvic arteries (sagittal section).

12d   Pelvic Vasculature (Coronal View)

The common iliac arteries divide at the sacroiliac joints and become the external and internal iliac arteries. The external iliac arteries mainly serve the lower limb. The internal iliac arteries distribute blood to the pelvic walls and viscera (i.e., rectum, bladder, prostate, ductus deferens, uterus, uterine tubes, and ovaries). In addition, these arteries distribute blood to the gluteal region, the perineum, and the medial compartment of the thigh.

A. COMMON ILIAC ARTERY.

B. EXTERNAL ILIAC ARTERY.

C. INTERNAL ILIAC ARTERY.

D. SUPERIOR GLUTEAL ARTERY.

E. INFERIOR GLUTEAL ARTERY.

F. OBTURATOR ARTERY.

G. SUPERIOR VESICAL ARTERY.

H. MIDDLE RECTAL ARTERY.

I. INTERNAL PUDENDAL ARTERY.

J. INFERIOR RECTAL ARTERY.

K. SUPERIOR RECTAL ARTERY.

L. MEDIAL SACRAL ARTERY.

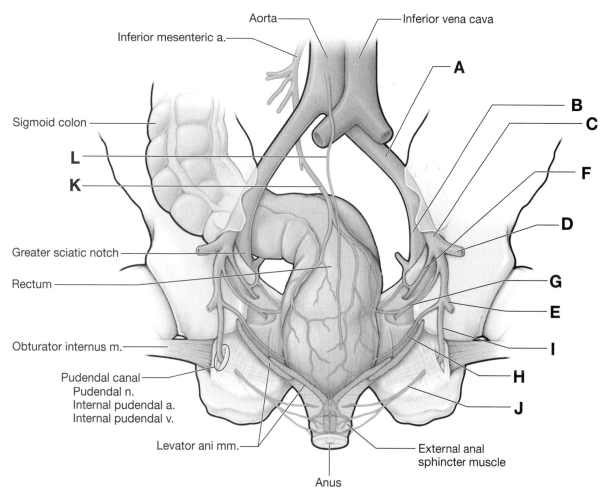

Aorta

Inferior vena cava

Inferior mesenteric a.

Sigmoid colon

A

B

C

L

F

K

D

Greater sciatic notch

G

Rectum

E

I

Obturator internus m.

H

Pudendal canal
Pudendal n.
Internal pudendal a.
Internal pudendal v.

J

Levator ani mm.

External anal
sphincter muscle

Anus

Pelvic arteries (coronal section from posterior view).

12e   Autonomic Innervation of Pelvis

Somatic and autonomic nerves contribute to pelvic innervation. The obturator nerve and the sacral plexus provide innervation to skeletal muscles and skin in the pelvis and lower limbs. All autonomics of the pelvis and perineum pass through the inferior hypogastric plexus. Sympathetic and parasympathetic nerves contribute to the inferior hypogastric plexus through the sacral splanchnics and pelvic splanchnics, respectively.

A. ABDOMINAL AORTA.

B. SYMPATHETIC GANGLIA AND TRUNK.

C. INFERIOR MESENTERIC GANGLION AND PLEXUS.

D. WHITE RAMUS COMMUNICANS.

E. GRAY RAMUS COMMUNICANS.

F. LUMBAR SPLANCHNIC NERVES.

G. SACRAL SPLANCHNIC NERVES.

H. PELVIC SPLANCHNIC NERVES.

I. SUPERIOR HYPOGASTRIC PLEXUS.

J. HYPOGASTRIC NERVES.

K. INFERIOR HYPOGASTRIC PLEXUS.

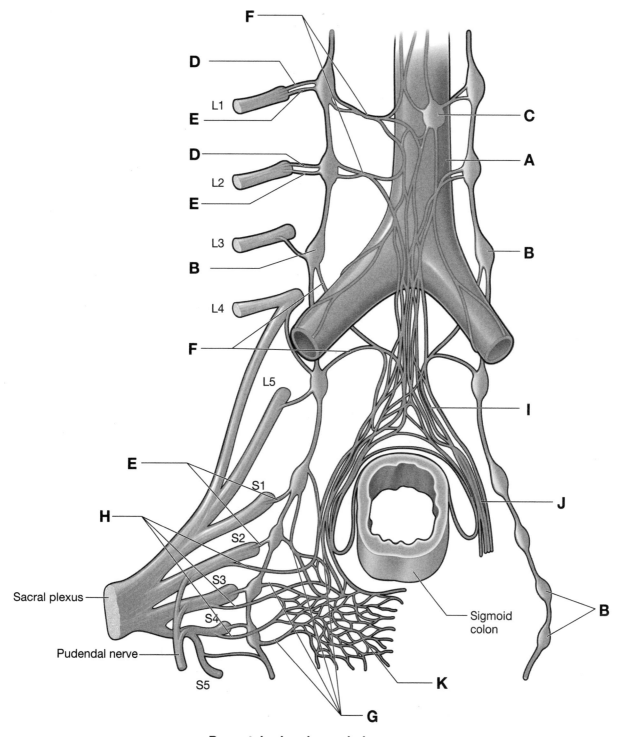

Prevertebral and sacral plexuses.

| 12f | **Veins of the Rectum** |

The colon terminates at the rectum, which in turn terminates at the anus. The rectum is located superior to the pelvic diaphragm in the pelvic cavity. The anus is located inferior to the pelvic diaphragm in the ischioanal fossa. The terminal end of the gastrointestinal tract is significant because it is a junction of two embryologic origins of tissue: endoderm and ectoderm.

Venous return is through the superior rectal vein (a branch of the inferior mesenteric vein of the portal system), the middle rectal vein (the internal iliac vein), and the inferior rectal vein (the internal pudendal vein).

A. RECTAL VENOUS PLEXUS.

B. SUPERIOR RECTAL VEIN.

C. INFERIOR MESENTERIC VEIN.

D. SPLENIC VEIN.

E. PORTAL VEIN.

F. MIDDLE RECTAL VEIN.

G. INFERIOR RECTAL VEIN.

H. INTERNAL PUDENDAL VEIN.

I. INTERNAL ILIAC VEIN.

J. COMMON ILIAC VEIN.

K. INFERIOR VENA CAVA.

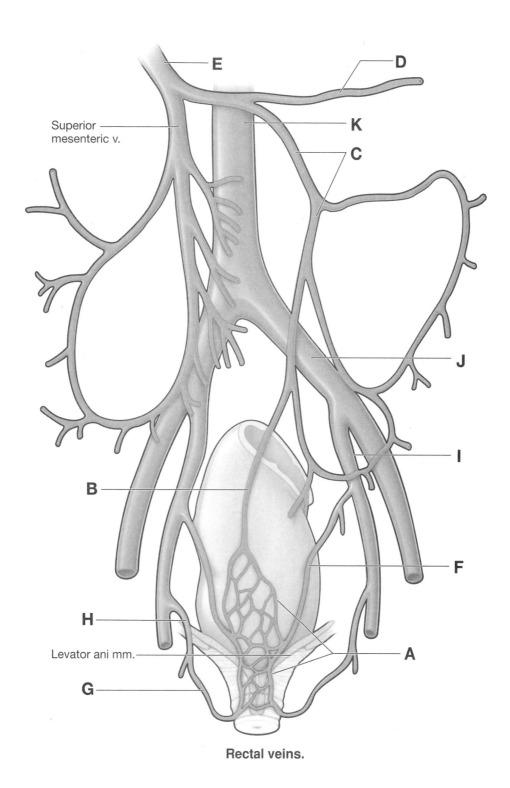

Rectal veins.

13.

Male Reproductive System

13a   Male Reproductive System (Sagittal Section)

The male reproductive system primarily consists of the paired testes and the penis. In addition, accessory sex glands contribute to seminal fluid. The male reproductive system matures during adolescence and remains active for the remainder of the lifespan of the male.

A. SCROTUM.

B. TESTIS.

C. EPIDIDYMIS.

D. DUCTUS DEFERENS.

E. URINARY BLADDER.

F. SEMINAL VESICLE.

G. EJACULATORY DUCT.

H. PROSTATE GLAND.

I. PROSTATIC URETHRA.

J. MEMBRANOUS URETHRA.

K. BULBOURETHRAL GLAND.

L. SPONGY URETHRA.

M. CORPORA CAVERNOSA.

N. CORPUS SPONGIOSUM.

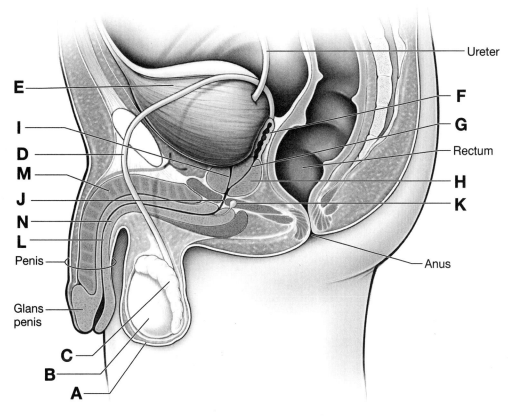

Male reproductive system.

13b   Penis

Each crus of the penis is composed of erectile tissue that is continuous with the paired corpus cavernosa of the penis. Erectile tissue consists of large, cavernous venous sinusoids (spaces like those found in a sponge) that normally are somewhat void of blood. However, during sexual arousal, the corpus cavernosa fill with blood, causing the penis to become erect. The deep arteries of the penis (branches of the internal pudendal artery) course within the center of the corpus cavernosa, providing blood that is necessary for an erection.

A. ISCHIOCAVERNOSUS MUSCLE.

B. BULBOSPONGIOSUS MUSCLE.

C. PERINEAL MEMBRANE.

D. DEEP TRANSVERSE PERINEAL MUSCLE.

E. CORPUS CAVERNOSA.

F. CORPUS SPONGIOSUM.

G. GLANS PENIS.

H. DEEP PENILE ARTERIES.

I. DEEP DORSAL PENILE VEIN.

J. DORSAL PENILE ARTERY.

K. DORSAL PENILE NERVE.

L. SUPERFICIAL DORSAL PENILE VEINS.

M. URETHRA.

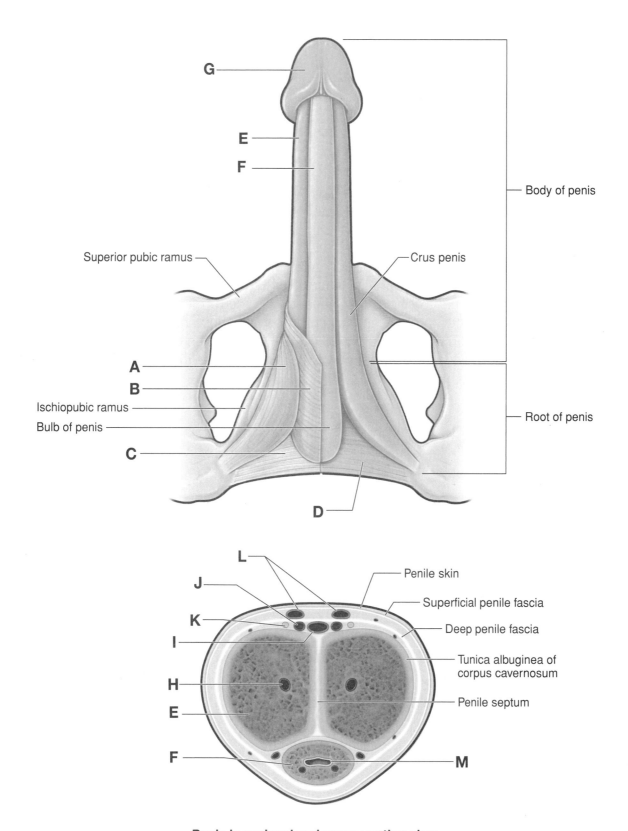

Penis in perineal and cross-section view.

14.

Female Reproductive System

14a	**Female Reproductive System (Sagittal Section)**

The female reproductive system consists of the ovaries, uterine tubes, uterus, vagina, and external genitalia. These organs remain underdeveloped for about the first 10 years of life. During adolescence, sexual development occurs and menses first occur (menarche). Cyclic changes occur throughout the reproductive period, with an average cycle length of approximately 28 days. These cycles cease at about the fifth decade of life (menopause), at which time the reproductive organs become atrophic.

A. UTERUS.

B. VAGINA.

C. BLADDER.

D. URETHRA.

E. PUBIC SYMPHYSIS.

F. CLITORIS.

G. RECTUM.

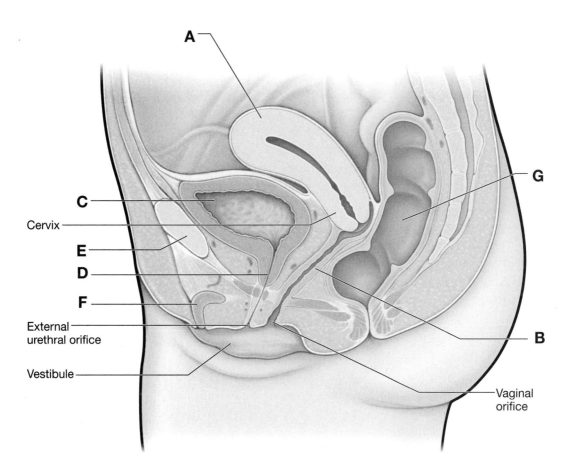

Female reproductive system (sagittal section).

14b   Female Reproductive System (Coronal Section)

The uterus, known as the womb, resembles an inverted pear and is located in the pelvic cavity between the rectum and the urinary bladder. The uterus is a hollow organ that functions to receive and nourish a fertilized oocyte until birth. Normally, the uterus is flexed anteriorly, where it joins the vagina; however, the uterus may also be retroverted (flexed posteriorly). The pelvic and urogenital diaphragms support the uterus.

The paired uterine tubes are also called fallopian tubes, or oviducts, and extend from the ovaries to the uterus. The luminal diameter of the uterine tubes is very narrow and, in fact, is only as wide as a human hair.

A. OVARY.

B. UTERINE TUBE.

C. MYOMETRIUM.

D. ENDOMETRIUM.

E. UTERINE CAVITY.

F. CERVICAL CANAL.

G. VAGINAL FORNIX.

H. CERVIX.

I. VAGINA.

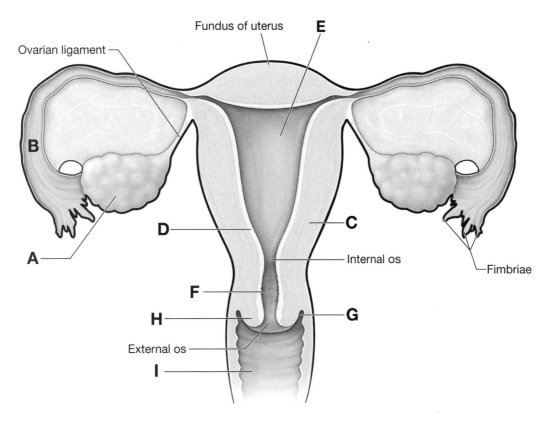

Coronal section of the uterus and uterine tubes.

14c	**Broad Ligament**

The uterine tubes and uterus are covered by a layer of peritoneum on the anterior, superior, and posterior surfaces. Inferior to the uterine tube and lateral to the uterus, the peritoneal membrane is fused into a double layer called the broad ligament.

A. UTERUS.

B. OVARY.

C. UTERINE TUBE.

D. BROAD LIGAMENT.

E. MESOSALPINX.

F. MESOVARIUM.

G. URETER.

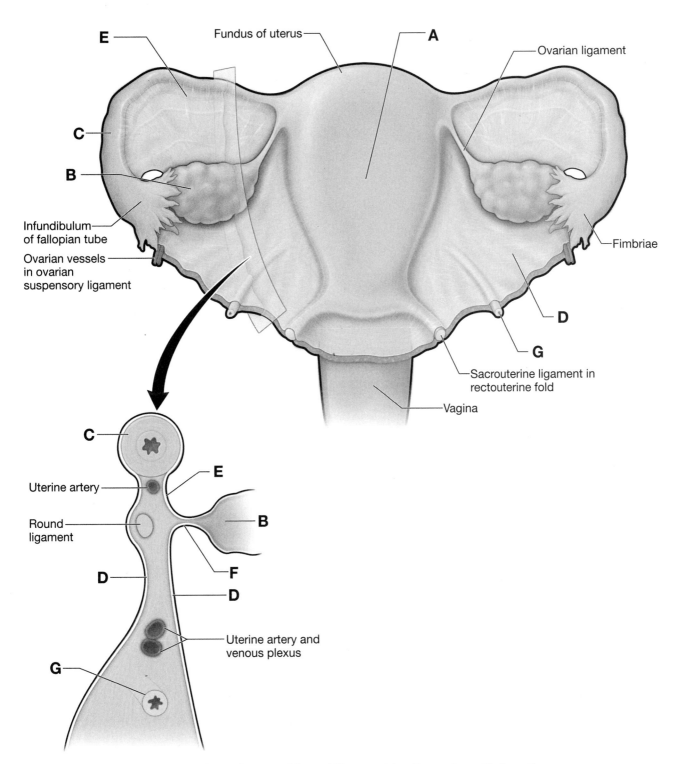

Uterus, uterine tubes, and broad ligament in situ and sagittal section.

SECTION IV

HEAD

15.

Scalp, Skull, and Meninges

15a   Scalp (Coronal Section)

The layers of the scalp can best be remembered by the acronym **"SCALP,"** with each letter of the word representing the tissue layer associated with it.

A. SKIN. Contains sweat glands, sebaceous glands, and hair follicles.

B. CONNECTIVE TISSUE. Composed of dense collagenous connective tissue as well as vessels and nerves.

C. APONEUROSIS. Consists of the frontalis muscle connected to the occipitalis muscle via the galea aponeurotica.

D. LOOSE CONNECTIVE TISSUE. A sponge-like layer of loose connective tissue.

E. PERICRANIUM. The periosteum on the external surface of the skull.

F. SKULL.

G. DURA MATER.

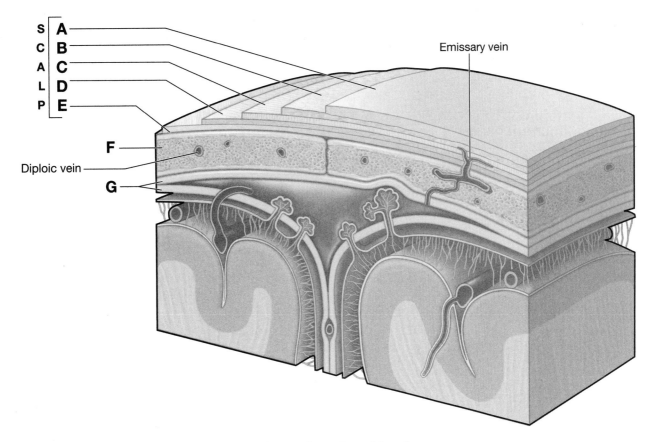

S
C
A
L
P

A
B
C
D
E

F

Diploic vein

G

Emissary vein

Coronal section of head.

15b   Scalp Innervation

The scalp receives its cutaneous innervation from a combination of branches from dorsal rami (greater occipital nerve), ventral rami (lesser occipital nerve), and cranial nerves (supraorbital and supra-trochlear nerves, zygomaticotemporal and auriculotemporal nerves).

A. AURICULOTEMPORAL NERVE (CN V-3).

B. GREATER OCCIPITAL NERVE (C2).

C. LESSER OCCIPITAL NERVE (C2).

D. ZYGOMATICOTEMPORAL NERVE (CN V-2).

E. SUPRAORBITAL NERVE (CN V-1).

F. SUPRATROCHLEAR NERVE (CN V-1).

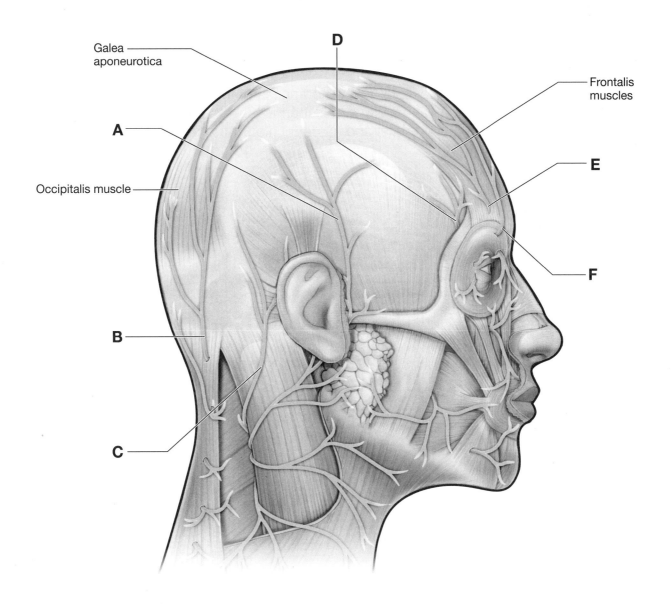

Cutaneous innervation of the scalp.

| 15c | **Scalp Arterial Supply** |

The scalp is highly vascularized via branches of the external carotid (occipital, posterior auricular, and superior temporal arteries) and internal carotid (supraorbital and supratrochlear) arteries.

A. SUPRAORBITAL ARTERY.

B. SUPRATROCHLEAR ARTERY.

C. SUPERFICIAL TEMPORAL ARTERY.

D. EXTERNAL CAROTID ARTERY.

E. COMMON CAROTID ARTERY.

F. INTERNAL CAROTID ARTERY.

G. OCCIPITAL ARTERY.

H. POSTERIOR AURICULAR ARTERY.

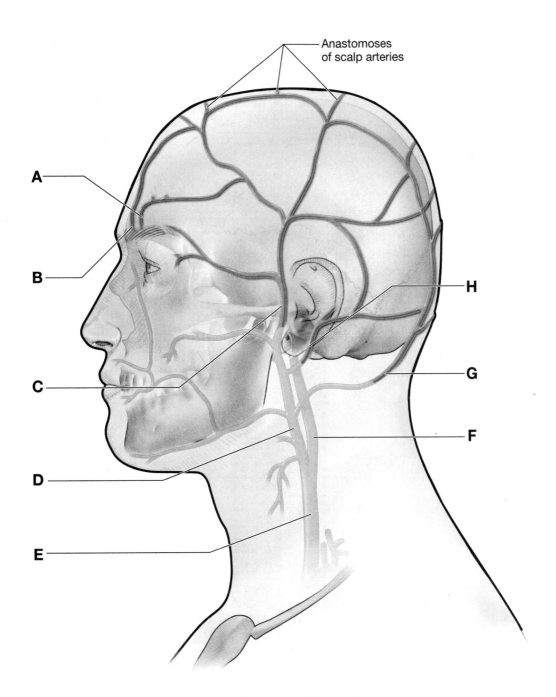

Anastomoses
of scalp arteries

Arterial supply of the scalp.

15d Skull

There are 8 cranial bones and 14 facial bones in the skull, all of which serve to protect the brain. The skull protects the brain and its surrounding meninges. The outer and inner surfaces of the skull are covered by periosteum, known, respectively, as the pericranium and the endocranium. The periosteum is continuous at the sutures of the skull. The cranial bones consist of spongy bone "sandwiched" between two layers of compact bone. Sutures are immovable fibrous joints between the bones of the skull. The principal bones and sutures of the skull are as follows:

A. FRONTAL BONE.

B. PARIETAL BONE.

C. CORONAL SUTURE.

D. TEMPORAL BONE.

E. PTERION.

F. SPHENOID BONE.

G. NASAL BONE.

H. LACRIMAL BONE.

I. ZYGOMATIC BONE.

J. MAXILLA.

K. MANDIBLE.

L. ZYGOMATIC ARCH.

M. SQUAMOUS SUTURE.

N. LAMBDOID SUTURE.

O. SAGITTAL SUTURE.

P. OCCIPITAL BONE.

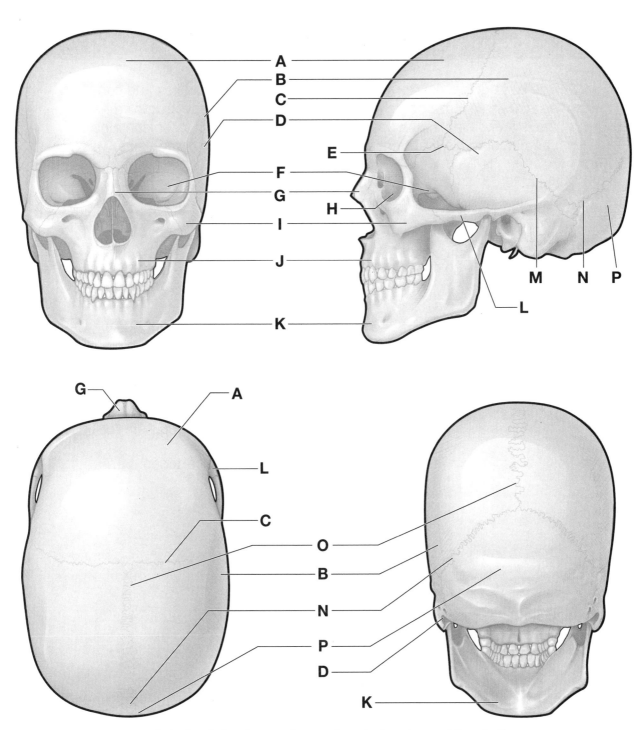

Anterior, lateral, superior, and posterior views of the skull.

15e   Base of the Skull (Superior View)

The base of the skull forms the floor on which the brain lies and consists of three large depressions that lie on different levels known as the anterior middle, and posterior cranial fossae. Each fossa contains various foramina that transmit cranial nerves, arteries, and veins.

A. CRIBRIFORM PLATE. CN I.

B. OPTIC NERVE. Transmits through optic canal.

C. SUPERIOR ORBITAL FISSURE. CNN III, IV, V-1, and VI.

D. CN V-2. Transmits through foramen rotundum.

E. CN V-3. Transmits through foramen ovale.

F. MIDDLE MENINGEAL ARTERY. Transmits foramen spinosum.

G. INTERNAL CAROTID ARTERY.

H. INTERNAL ACOUSTIC MEATUS. CNN VII and VIII.

I. JUGULAR FORAMEN. CNN IX, X, and XI.

J. VERTEBRAL ARTERY.

K. FORAMEN MAGNUM. Spinal cord and vertebral arteries.

L. SPINAL CORD.

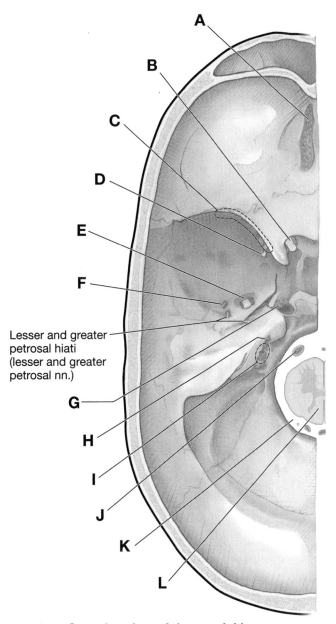

Superior view of the cranial base.

15f   Base of the Skull (Inferior View)

The base of the skull forms the floor on which the brain lies and consists of three large depressions that lie on different levels known as the anterior middle, and posterior cranial fossae. Each fossa contains various foramina that transmit cranial nerves, arteries, and veins.

A. INCISIVE CANAL. Nasopalatine nerve and artery.

B. GREATER AND LESSER PALATINE NERVES.

C. FORAMEN OVALE. CN V-3.

D. FORAMEN SPINOSUM. Transmits middle meningeal artery.

E. INTERNAL CAROTID ARTERY.

F. STYLOMASTOID FORAMEN. CN VII.

G. JUGULAR FORAMEN. CNN IX, X, XI and internal jugular vein.

H. VERTEBRAL ARTERY.

I. FORAMEN MAGNUM. Spinal cord and vertebral arteries.

J. SPINAL CORD.

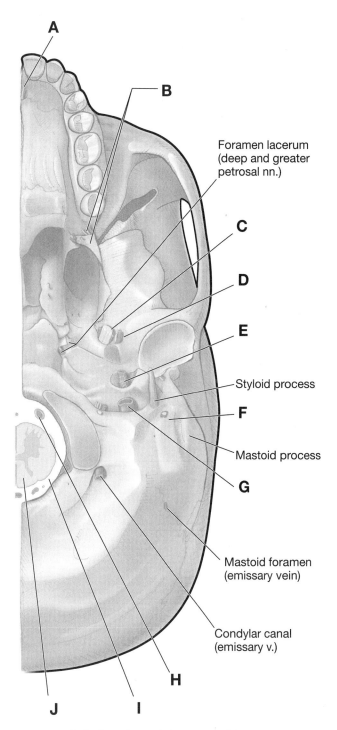

A

B

Foramen lacerum
(deep and greater
petrosal nn.)

C

D

E

Styloid process

F

Mastoid process

G

Mastoid foramen
(emissary vein)

Condylar canal
(emissary v.)

H

J   **I**

Inferior view of the cranial base.

15g   Meninges (Coronal Section)

The brain is surrounded and protected by three connective tissue layers called meninges. These meninges, from superficial to deep, are the dura mater, arachnoid mater, and pia mater.

A. SCALP.

B. SKULL.

C. DURA MATER.

D. FALX CEREBRI.

E. PERIOSTEAL LAYER OF DURA MATER.

F. MENINGEAL LAYER OF DURA MATER.

G. SUPERIOR SAGITTAL SINUS.

H. ARACHNOID MATER.

I. SUBARACHNOID SPACE.

J. ARACHNOID GRANULATION.

K. PIA MATER.

L. GRAY MATTER.

M. WHITE MATTER.

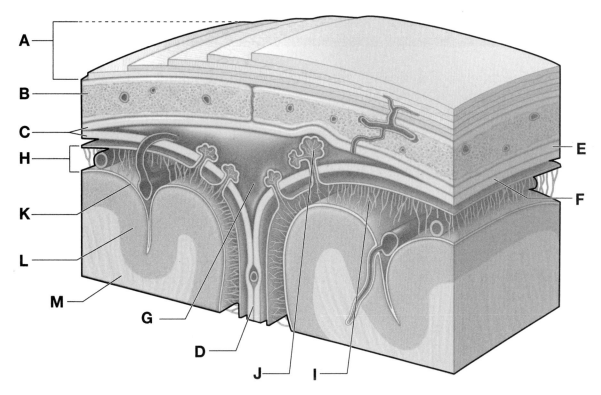

Cranial meninges (coronal section).

15h   Dural Venous Sinuses

The dural venous sinuses are venous channels located between the periosteal and meningeal layers of the dura mater. The dural venous sinuses are lined with endothelium and lack valves. They serve as a receptacle for blood from the cerebral, diploic, and emissary veins. They also receive the cerebrospinal fluid (CSF), drained by the arachnoid granulations. Blood in the dural venous sinuses primarily drains into the internal jugular veins.

A. SUPERIOR SAGITTAL SINUS.

B. INFERIOR SAGITTAL SINUS.

C. STRAIGHT SINUS.

D. CONFLUENCE OF SINUSES.

E. TRANSVERSE SINUS.

F. SIGMOID SINUS.

G. INFERIOR PETROSAL SINUS.

H. SUPERIOR PETROSAL SINUS.

I. CAVERNOUS SINUS.

J. INTERNAL JUGULAR VEIN.

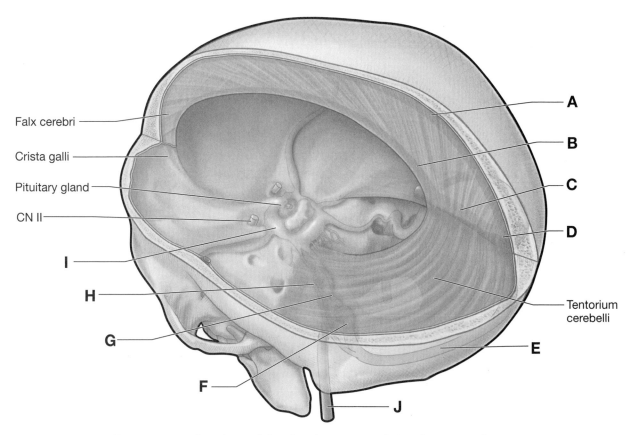

Falx cerebri

Crista galli

Pituitary gland

CN II

I

H

G

F

A

B

C

D

Tentorium cerebelli

E

J

Posterosuperior view of the dural septae and dural venous sinuses.

Section IV · Head

15i   Cavernous Sinus (Coronal Section)

The cavernous sinus is a dural venous sinus located on each side of the sella turcica. The internal carotid artery and CN VI course through the middle of the sinus. Cranial nerves (CNN) III, IV, V-1, and V-2 course through the lateral walls of the sinus. The cavernous sinus communicates with the pterygoid venous plexus via emissary veins and the superior and inferior ophthalmic veins.

A. CN II.

B. CN III.

C. CN IV.

D. CN VI.

E. CN V-1.

F. CN V-2.

G. INTERNAL CAROTID ARTERY (CEREBRAL PART).

H. INTERNAL CAROTID ARTERY (CAVERNOUS PART).

I. PITUITARY GLAND.

J. CAVERNOUS SINUS.

K. SPHENOID SINUS.

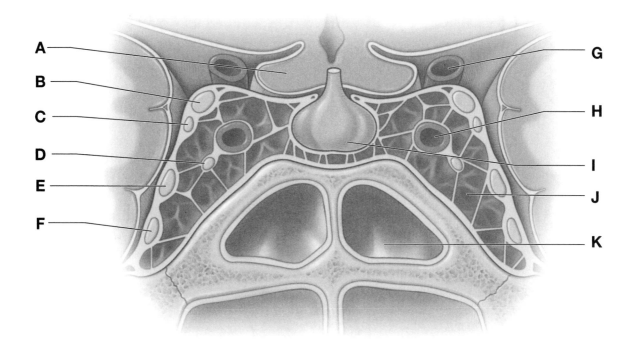

Coronal section through the sphenoid bone highlighting the cavernous sinuses.

16.

Brain

| 16a | **Brain** |

The brain contains millions of neurons arranged in a vast array of synaptic connections that provide seemingly unfathomable circuitry. Through that circuitry, the brain integrates and processes sensory information and provides motor output. The brain is divided into the cerebrum, diencephalon (epithalamus, thalamus, hypothalamus), brainstem (midbrain, pons, medulla) and cerebellum.

A. FRONTAL LOBE.

B. PARIETAL LOBE.

C. OCCIPITAL LOBE.

D. TEMPORAL LOBE.

E. CEREBELLUM.

F. MIDBRAIN.

G. PONS.

H. MEDULLA OBLONGATA.

I. SPINAL CORD.

J. EPITHALAMUS.

K. THALAMUS.

L. HYPOTHALAMUS.

M. PITUITARY GLAND.

N. CORPUS CALLOSUM.

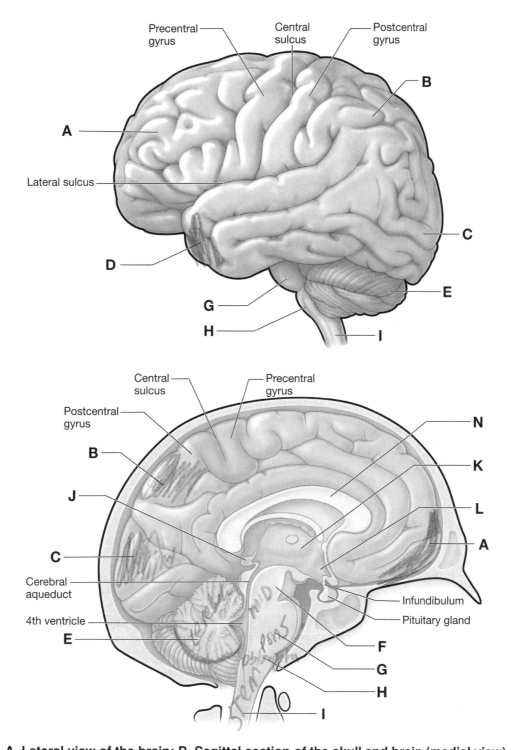

A. Lateral view of the brain; B. Sagittal section of the skull and brain (medial view).

| 16b | **Ventricles** |

The ventricular system of the brain is a set of four chambers within the brain and is continuous with the central canal of the spinal cord. Cerebrospinal fluid (CSF) flows within these chambers and serves as a liquid cushion, providing buoyancy to the brain and spinal cord.

A. LATERAL VENTRICLES.

B. 3RD VENTRICLE.

C. CEREBRAL AQUEDUCT.

D. 4TH VENTRICLE.

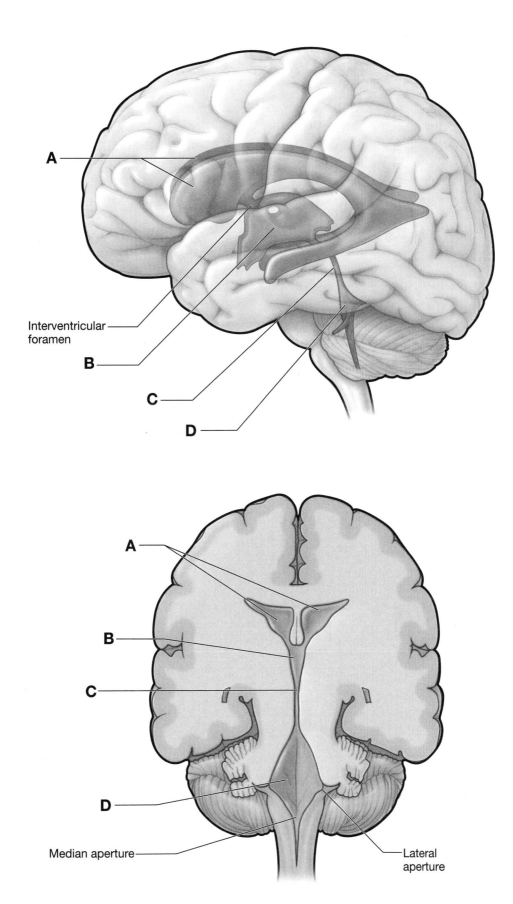

Interventricular foramen

Median aperture

Lateral aperture

A. Lateral view of ventricles; B. Coronal section of ventricles.

16c　Cerebrospinal Fluid (CSF)

CSF is produced by the choroid plexuses, located within each of the ventricles, and flows from the lateral and third ventricles to the fourth ventricle via the cerebral aqueduct. From the fourth ventricle, CSF enters an enlarged part of the subarachnoid space (cisterna magna) via the central median aperture (of Magendie) and the lateral apertures (of Luschka). The CSF circulates around the spinal cord and brain in the subarachnoid space to empty into the superior sagittal sinus via the arachnoid granulations. Arachnoid granulations are projections of the arachnoid mater along the superior sagittal sinus.

A. CHOROID PLEXUS.

B. LATERAL VENTRICLE.

C. 3RD VENTRICLE.

D. CEREBRAL AQUEDUCT.

E. 4TH VENTRICLE.

F. SUBARACHNOID SPACE.

G. ARACHNOID GRANULATIONS.

H. SUPERIOR SAGITTAL SINUS.

I. CONFLUENCE OF SINUSES.

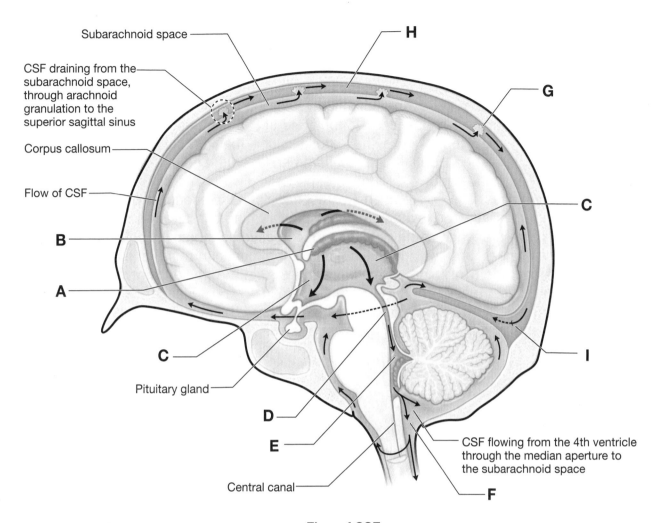

Subarachnoid space

CSF draining from the subarachnoid space, through arachnoid granulation to the superior sagittal sinus

Corpus callosum

Flow of CSF

H

G

C

B

A

C

Pituitary gland

D

E

Central canal

CSF flowing from the 4th ventricle through the median aperture to the subarachnoid space

F

I

Flow of CSF.

16d   Arterial Supply of the Brain

The brain receives its arterial supply from branches of the following two sources, the internal carotid and the vertebral arteries. The internal carotid artery ascends to the base of the skull, entering the carotid canal. Upon exiting the carotid canal, the internal carotid artery courses horizontally over the foramen lacerum and enters the cavernous sinus and, after turning superiorly, divides into its terminal branches. The vertebral arteries course along the inferior aspect of the medulla oblongata before converging into the basilar artery on the pons. Branches of the vertebral arteries travel to the spinal cord, the meninges, and the brainstem. The anterior communicating artery connects the two anterior cerebral arteries, and the posterior communicating arteries connect the internal carotid and posterior cerebral arteries. As a result of these connections, an arterial circle, known as the cerebral arterial circle (of Willis), is formed.

A. INTERNAL CAROTID ARTERY.

B. ANTERIOR CEREBRAL ARTERY.

C. ANTERIOR COMMUNICATING ARTERY.

D. MIDDLE CEREBRAL ARTERY.

E. POSTERIOR COMMUNICATING ARTERY.

F. VERTEBRAL ARTERY.

G. BASILAR ARTERY.

H. SUPERIOR CEREBELLAR ARTERY.

I. POSTERIOR CEREBRAL ARTERY.

J. ANTERIOR INFERIOR CEREBELLAR ARTERY.

K. POSTERIOR INFERIOR CEREBELLAR ARTERY.

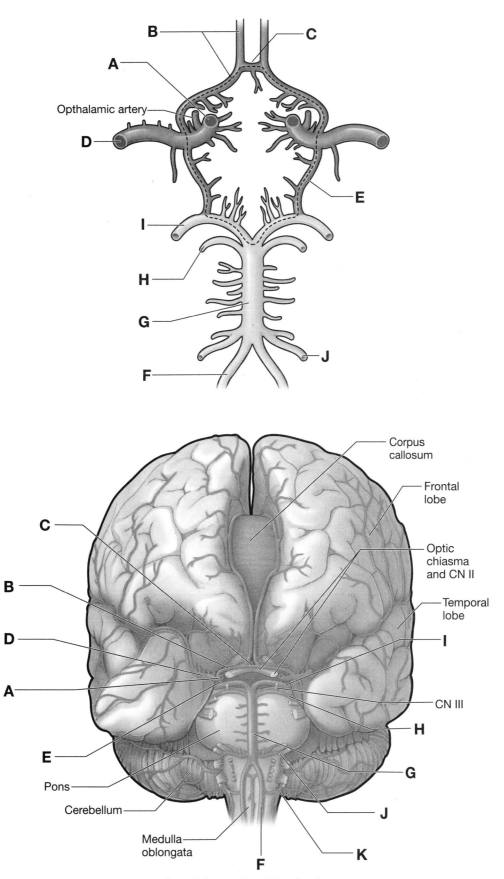

Opthalamic artery

Corpus callosum

Frontal lobe

Optic chiasma and CN II

Temporal lobe

CN III

Pons

Cerebellum

Medulla oblongata

Arterial supply of the brain.

17.

Cranial Nerves

| 17a | **Cranial Nerves (Overview)** |

Cranial nerves emerge through openings in the skull and are covered by tubular sheaths of connective tissue derived from the cranial meninges. There are 12 pairs of cranial nerves, numbered I to XII, from rostral to caudal, according to their attachment to the brain. The names of the cranial nerves reflect their general distribution and function. Like spinal nerves, cranial nerves are bundles of sensory and motor neurons that conduct impulses from sensory receptors and innervate muscles or glands.

A. SOMATIC MOTOR NEURONS.

B. BRANCHIAL MOTOR NEURONS.

C. VISCERAL MOTOR (PARASYMP) NEURONS.

D. GENERAL SENSORY NEURONS.

E. SPECIAL SENSORY NEURONS.

F. VISCERAL SENSORY NEURONS.

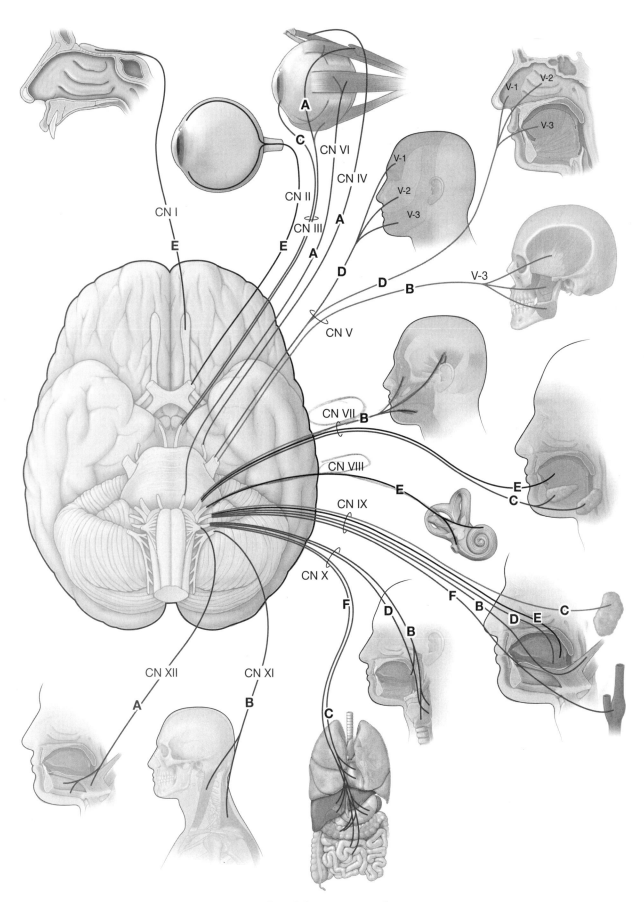

Cranial nerve overview.

17b   CN I: Olfactory Nerve and CN II: Optic Nerve

The olfactory neurons originate in the olfactory epithelium in the superior part of the lateral and septal walls of the nasal cavity. The nerves ascend through the cribriform foramina of the ethmoid bone to reach the olfactory bulbs. The olfactory neurons synapse with neurons in the bulbs, which course to the primary and association areas of the cerebral cortex.

Optic nerve fibers arise from the retina and all converge at the optic disc. CN II exits the orbit via the optic canals. Both optic nerves form the optic chiasm, the site where neurons from the nasal side of either retina cross over to the contralateral side of the brain. The neurons then pass via the optic tracts to the thalamus, where they synapse with neurons that course to the primary visual cortex of the occipital lobe.

A. OLFACTORY NERVES.

B. OLFACTORY BULB.

C. OLFACTORY TRACT.

D. OPTIC NERVE.

E. OPTIC CHIASM.

F. OPTIC TRACT.

G. LATERAL GENICULATE NUCLEUS.

H. OPTIC RADIATIONS.

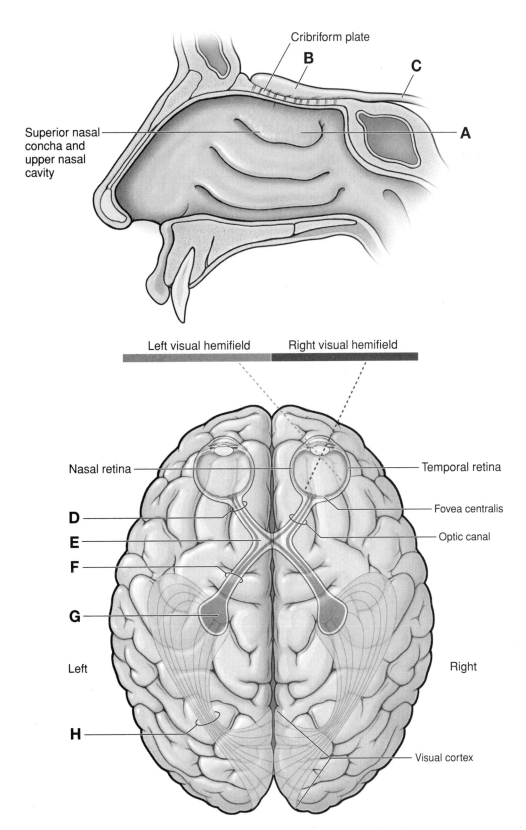

Cribriform plate

B

C

Superior nasal concha and upper nasal cavity

A

Left visual hemifield

Right visual hemifield

Nasal retina

Temporal retina

D

Fovea centralis

E

F

Optic canal

G

Left

Right

H

Visual cortex

A: Olfactory nerve (CN I). B: Optic nerve (CN II).

17c CN III: Oculomotor Nerve, CN IV: Trochlear Nerve, and CN VI: Abducens Nerve

The oculomotor nerve innervates the levator palpebrae superioris muscle, four of the six extraocular muscles (superior rectus, medial rectus, inferior rectus and inferior oblique) as well as the pupillary constrictor and ciliary muscles.

The trochlear nerve innervates the superior oblique muscle.

The abducens nerve innervates the lateral rectus muscle.

A. CN III: OCULOMOTOR NERVE.

B. LEVATOR PALPEBRAE SUPERIORIS MUSCLE.

C. SUPERIOR RECTUS MUSCLE.

D. INFERIOR RECTUS MUSCLE.

E. INFERIOR OBLIQUE MUSCLE.

F. MEDIAL RECTUS MUSCLE.

G. CILIARY MUSCLE (LENS ACCOMMODATION).

H. CILIARY GANGLION.

I. SPHINCTER PUPILLAE MUSCLE (CONSTRICT PUPIL).

J. CN IV: TROCHLEAR NERVE.

K. SUPERIOR OBLIQUE MUSCLE.

L. CN VI: ABDUCENS NERVE.

M. LATERAL RECTUS MUSCLE.

N. INTERNAL CAROTID ARTERY AND PLEXUS.

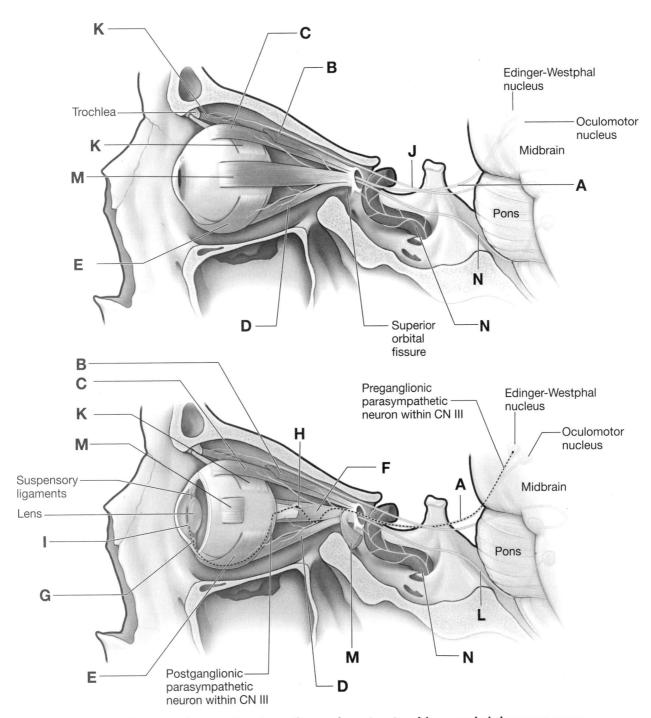

A: Somatic motor innervation from the oculomotor, trochlear, and abducens nerves (CNN III, IV, and VI, respectively). B: Visceral motor parasympathetic component of CN III.

17d CN V: Trigeminal Nerve

The trigeminal nerve is the principal general sensory supply to the head. CN V originates from the lateral surface of the pons as a large sensory root and a smaller motor root. These roots enter the trigeminal (Meckel's) cave of the dura, lateral to the body of the sphenoid bone and the cavernous sinus. The sensory root leads to the trigeminal (semilunar) ganglion, which houses the cell bodies for the general sensory neurons. The motor root runs parallel to the sensory root, bypassing the ganglion and becoming part of the mandibular nerve (CN V-3). As well as being the primary general sensory distribution to the head, CN V also aids in distributing postganglionic parasympathetic neurons of the head to their destinations for CNN III, VII, and IX. The trigeminal ganglion gives rise to three divisions, named for the cranial location to the eyes (ophthalmic), the maxilla (maxillary), and the mandible (mandibular).

A. CN V: TRIGEMINAL NERVE ROOT.

B. TRIGEMINAL GANGLION.

C. CN V-1. Ophthalmic branch and distribution.

D. CN V-2. Maxillary branch and distribution.

E. CN V-3. Mandibular branch and distribution.

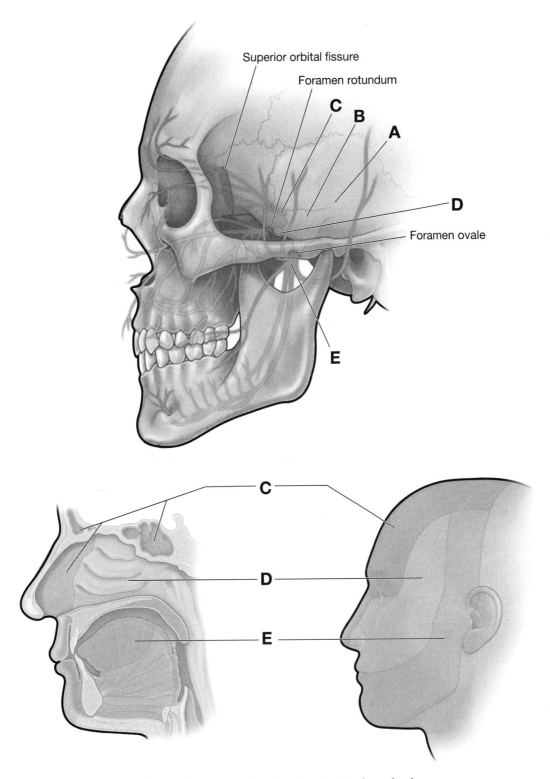

Superior orbital fissure

Foramen rotundum

Foramen ovale

A: General sensory innervation from trigeminal nerve.
B: General sensory distribution of CN V.

17e   CN VII: Facial Nerve

The facial nerve provides motor innervation to the muscles of facial expression, lacrimal gland, and submandibular and sublingual salivary glands, as well as taste to the anterior two-thirds of the tongue. Two distinct fascial sheaths package the four modalities carried by CN VII, with branchial motor neurons in one sheath and visceral motor, special sensory, and general sensory neurons in another sheath called the nervus intermedius. The nervous intermedius gives rise to the greater petrosal and chorda tympani nerves.

A. CN VII: FACIAL NERVE ROOT.

B. GENICULATE GANGLION.

C. FACIAL NERVE PROPER.

D. TEMPORAL BRANCH.

E. ZYGOMATIC BRANCH.

F. BUCCAL BRANCHES.

G. MANDIBULAR BRANCH.

H. CERVICAL BRANCH.

I. CHORDA TYMPANI NERVE.

J. SUBMANDIBULAR GANGLION.

K. GREATER PETROSAL NERVE.

L. PTERYGOPALATINE GANGLION.

M. CN V-3: MANDIBULAR BRANCH.

N. LINGUAL NERVE.

O. INFERIOR ALVEOLAR NERVE.

P. CN V-2: MAXILLARY BRANCH.

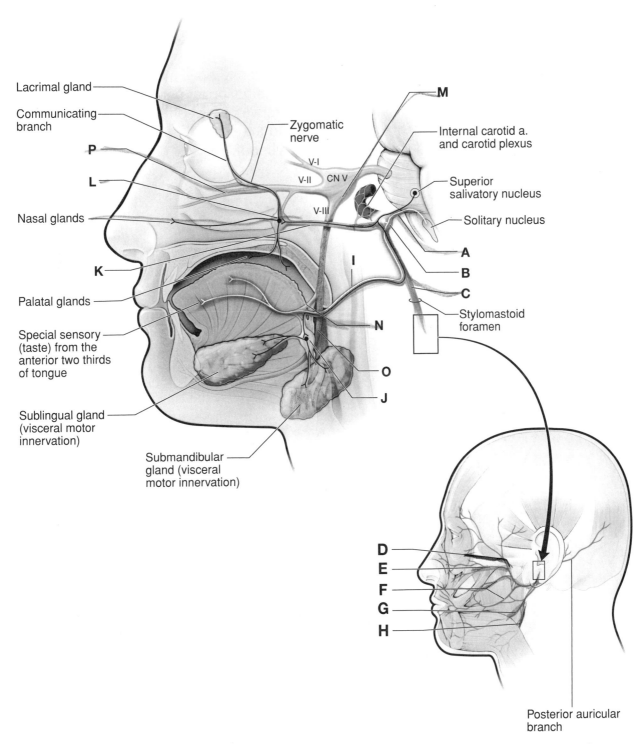

Lacrimal gland

Communicating branch

P

L

Nasal glands

K

Palatal glands

Special sensory (taste) from the anterior two thirds of tongue

Sublingual gland (visceral motor innervation)

Submandibular gland (visceral motor innervation)

Zygomatic nerve

V-I

V-II   CN V

V-III

M

Internal carotid a. and carotid plexus

Superior salivatory nucleus

Solitary nucleus

A

B

C

Stylomastoid foramen

I

N

O

J

D
E
F
G
H

Posterior auricular branch

A: The facial nerve (CN VII). B: Branchial motor innervation of superficial facial muscles.

17f   CN VIII: Vestibulocochlear Nerve

The vestibulocochlear nerve traverses the internal acoustic meatus with CN VII and provides special sensory innervation for hearing and equilibrium. CN VIII originates from the grooves between the pons and the medulla oblongata and divides into the cochlear branch for hearing and the vestibular branch for equilibrium.

A. CN VIII: VESTIBULOCOCHLEAR NERVE.

B. COCHLEA (HEARING).

C. SEMICIRCULAR CANALS (EQUILIBRIUM).

D. EAR OSSICLES.

E. TYMPANIC MEMBRANE.

F. AUDITORY TUBE.

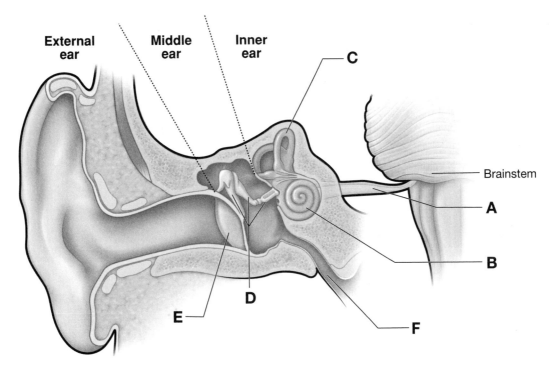

External ear Middle ear Inner ear

C

Brainstem

A

B

D

E

F

Special sensory distribution of the vestibulocochlear nerve (CN VIII).

17g   CN IX: Glossopharyngeal Nerve

The glossopharyngeal nerve provides motor innervation to the stylo-pharyngeus muscle and parotid gland and sensory innervation from the carotid body and sinus, posterior one-third of the tongue, and the auditory tube.

A. CN IX: GLOSSOPHARYNGEAL NERVE.

B. GLOSSOPHARYNGEAL GANGLIA.

C. CAROTID BRANCH OF CN IX.

D. PHARYNGEAL BRANCH OF CN IX.

E. BRANCHIAL MOTOR BRANCH OF CN IX.

F. TYMPANIC PLEXUS AND NERVE.

G. LESSER PETROSAL NERVE.

H. OTIC GANGLION.

I. AURICULOTEMPORAL NERVE.

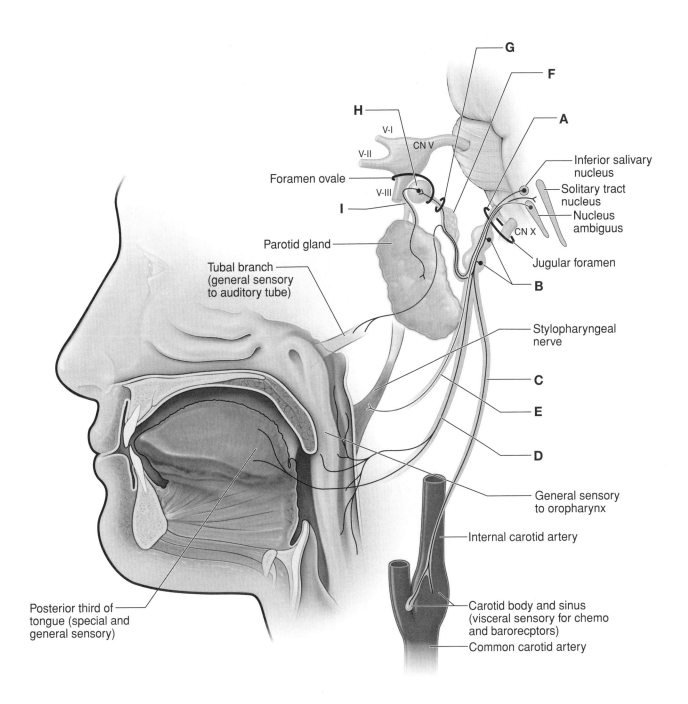

The glossopharyngeal nerve (CN IX).

17h CN X: Vagus Nerve

The vagus nerve innervates muscles of the larynx, pharynx, palate in addition to the gut tube, heart, and lungs.

The vagus nerve emerges from the lateral aspect of the medulla oblongata and traverses the jugular foramen, where the superior and inferior sensory ganglia are located. CN X gives rise to the pharyngeal (pharyngeal muscles), superior laryngeal (external and internal branches), recurrent laryngeal and visceral (parasympathetic) branches.

A. CN X: VAGUS NERVE.

B. VAGAL GANGLIA.

C. PHARYNGEAL BRANCH OF CN X.

D. CAROTID BODY BRANCH OF CN X.

E. SUPERIOR LARYNGEAL NERVE.

F. PALATAL BRANCH OF CN X.

G. EXTERNAL LARYNGEAL NERVE.

H. LEFT RECURRENT LARYNGEAL NERVE.

I. INTERNAL LARYNGEAL NERVE.

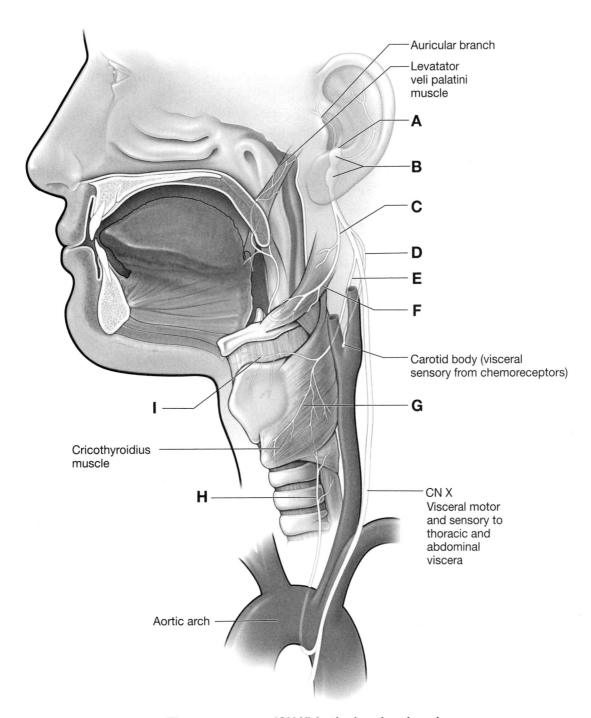

Auricular branch

Levatator veli palatini muscle

A

B

C

D

E

F

Carotid body (visceral sensory from chemoreceptors)

G

Cricothyroidius muscle

I

H

CN X Visceral motor and sensory to thoracic and abdominal viscera

Aortic arch

The vagus nerve (CN X) in the head and neck.

17i   CN XI: Spinal Accessory Nerve and CN XII: Hypoglossal Nerve

The spinal accessory nerve originates from the medulla oblongata, exits the jugular foramen, and provides branchial motor innervation to the trapezius and sternocleidomastoid muscles.

The hypoglossal nerve exits the medulla oblongata in the groove between the pyramid and the olive. Upon exiting the hypoglossal canal, CN XII courses through the neck, along the lateral surface of the hyoglossus muscle deep to the mylohyoid muscle and provides somatic motor innervation to intrinsic and extrinsic tongue muscles (except the palatoglossus, which is supplied by CN X).

A. SPINAL ROOT OF CN XI.

B. TRAPEZIUS MUSCLE.

C. STERNOCLEIDOMASTOID MUSCLE.

D. HYPOGLOSSAL CANAL.

E. STYLOGLOSSUS MUSCLE.

F. HYOGLOSSUS MUSCLE.

G. GENIOGLOSSUS MUSCLE.

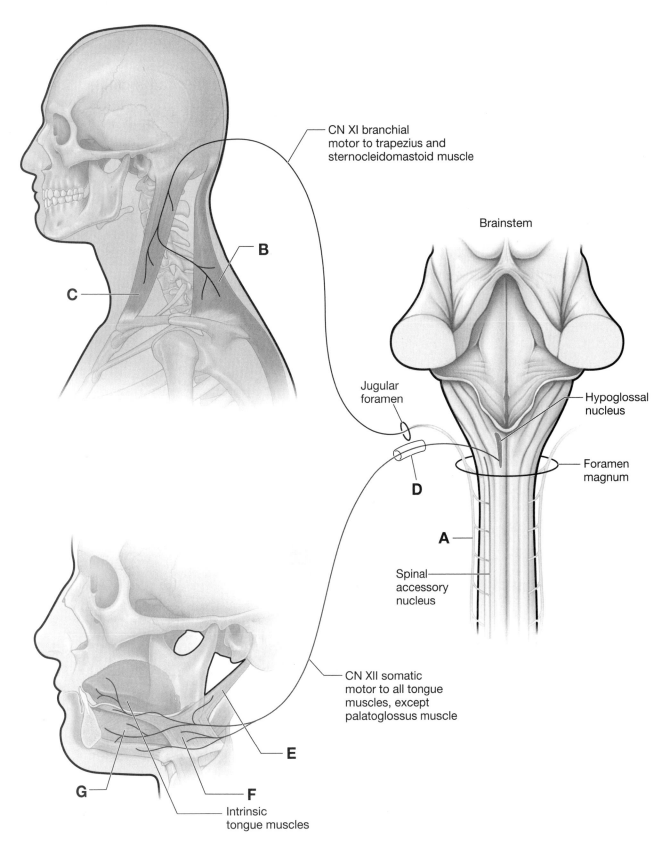

The spinal accessory nerve (CN XI) and hypoglossal nerve (CN XII).

17j   Autonomic Innervation of the Head

All preganglionic sympathetic neurons destined for the head originate at the T1 level of the spinal cord and synapse in the superior cervical ganglion. Postganglionic sympathetic neurons course along cranial arteries to the end organs such as the sweat glands, the superior tarsal muscle, and the dilator pupillae muscle.

Preganglionic parasympathetic neurons originate in the brainstem, course in CNN III, VII, IX, or X, and synapse in one of four ganglia (i.e., ciliary, pterygopalatine, submandibular, and otic). Postganglionic parasympathetic neurons then course along nerves to their end organs (e.g., salivary glands and pupillary sphincter muscle).

A. T1 LEVEL OF SPINAL CORD.

B. VENTRAL ROOT.

C. WHITE RAMUS COMMUNICANS.

D. SYMPATHETIC GANGLIA.

E. SYMPATHETIC TRUNK.

F. CN III: OCULOMOTOR NERVE.

G. CILIARY GANGLION.

H. CN VII: FACIAL NERVE.

I. GREATER PETROSAL NERVE.

J. PTERYGOPALATINE GANGLION.

K. CHORDA TYMPANI NERVE.

L. SUBMANDIBULAR GANGLION.

M. CN IX: GLOSSOPHARYNGEAL NERVE.

N. OTIC GANGLION.

O. CN X: VAGUS NERVE.

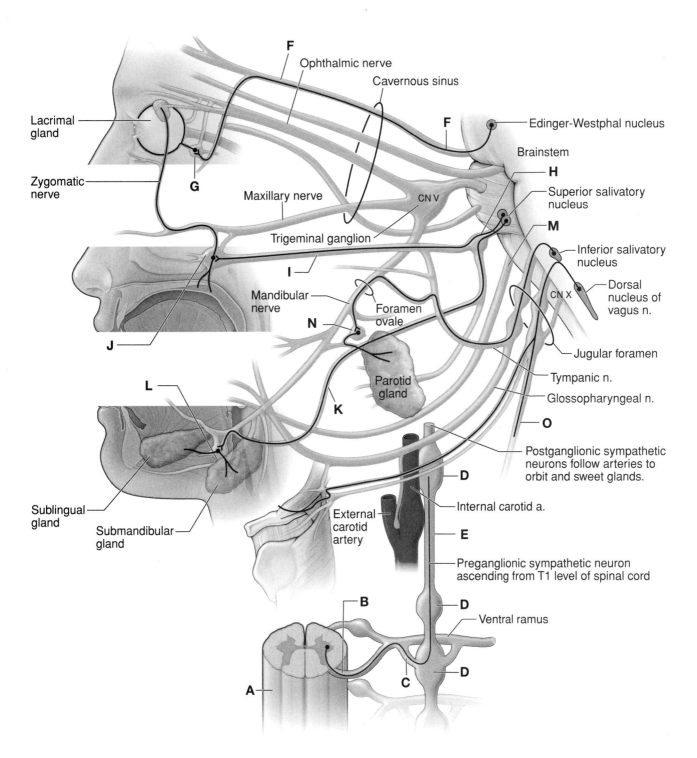

Autonomic innervation of the head.

18.

Orbit

| 18a | **Bony Orbit** |

The bony orbit is the region of the skull that surrounds the eye and is composed of the following structures:

A. FRONTAL BONE.

B. ETHMOID BONE.

C. LACRIMAL BONE.

D. PALATINE BONE.

E. MAXILLA.

F. ZYGOMATIC BONE.

G. SPHENOID BONE.

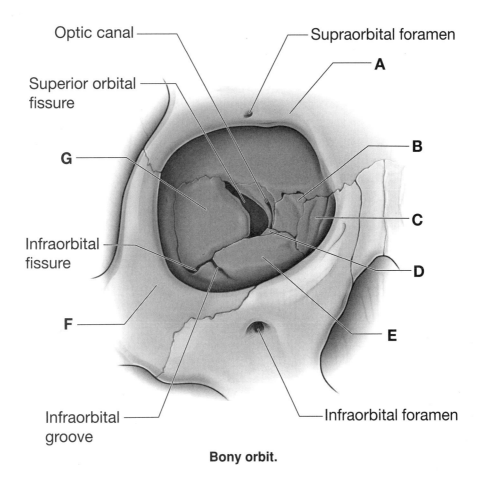

Optic canal

Supraorbital foramen

A

Superior orbital fissure

G

B

C

Infraorbital fissure

D

F

E

Infraorbital groove

Infraorbital foramen

Bony orbit.

18b　Eyelids

The primary organ responsible for vision is the eye. The eyeball is located within a bony orbital encasement, which protects it. The lacrimal apparatus keeps the eye moist and free of dust and other irritating particles through the production and drainage of tears. Eyelids protect the eye from external stimuli such as dust, wind, and excessive light. The external surface of the eyelids is covered by skin, whereas the conjunctiva covers the internal surface.

A. LEVATOR PALPEBRAE SUPERIORIS MUSCLE.

B. ORBICULARIS OCULI MUSCLE.

C. SUPERIOR TARSAL MUSCLE.

D. CONJUNCTIVA.

E. CORNEA.

F. IRIS.

G. CILIARY MUSCLES.

H. LENS.

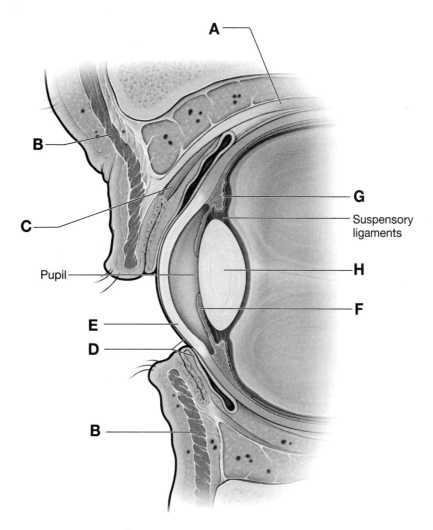

Pupil

Suspensory ligaments

Sagittal section of the eyelid.

18c   Eye

The eyeball consists of the sclera, choroid, and retina. The sclera is the white, fibrous covering of the eye into which extraocular muscles insert. The choroid is the vascular, middle layer containing the ciliary apparatus responsible for lens accommodation. The retina is the innermost layer of the eyeball consisting of rods and cones.

A. CONJUNCTIVA.

B. SCLERA.

C. CORNEA.

D. CHOROID.

E. CILIARY MUSCLE.

F. IRIS.

G. LENS.

H. RETINA.

I. CN II: OPTIC NERVE.

J. ANTERIOR CHAMBER.

K. POSTERIOR CHAMBER.

L. VITREOUS CHAMBER.

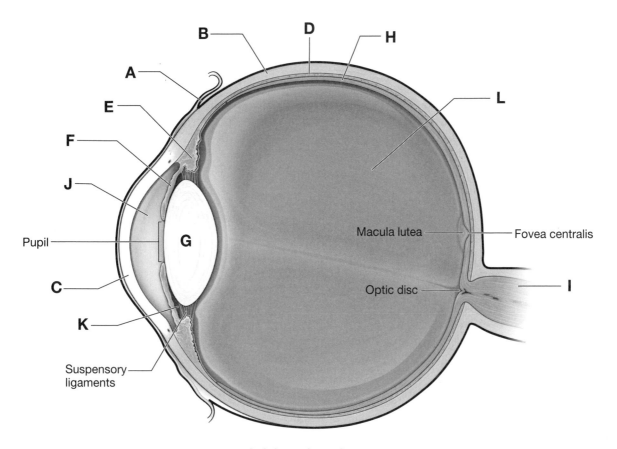

Axial section of eye.

18d   Extraocular Muscles

Six strap-like extraocular muscles (four rectus and two oblique) control the movement of the eye (elevation, depression, abduction, adduction, intorsion, and extorsion). These muscles are innervated by CNN III, IV, and VI.

A. LEVATOR PALPEBRAE SUPERIORIS MUSCLE.

B. SUPERIOR RECTUS MUSCLE.

C. SUPERIOR OBLIQUE MUSCLE.

D. TROCHLEA.

E. MEDIAL RECTUS MUSCLE.

F. LATERAL RECTUS MUSCLE.

G. INFERIOR RECTUS MUSCLE.

H. INFERIOR OBLIQUE MUSCLE.

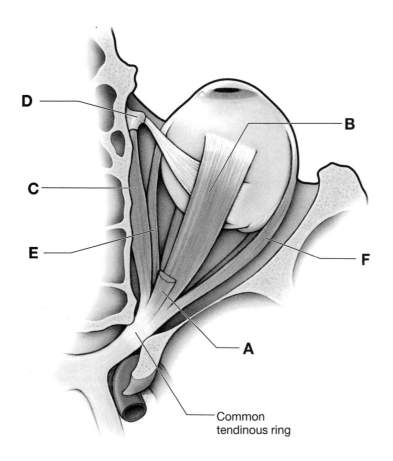

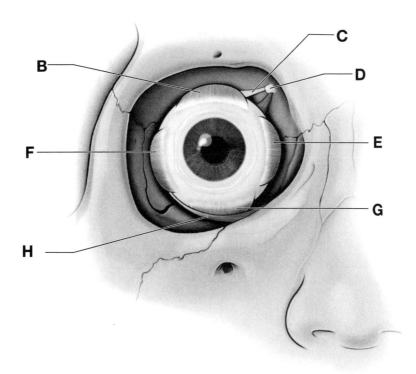

Superior and anterior views of extraocular muscles.

18e   Innervation of the Orbit

The cranial nerves associated with the orbit are CN II (vision), CN III (eye movement), CN IV (eye movement), CN V-1 (general sensory to eye and scalp), CN VI (eye movement), and CN VII (crying and closing the eyes).

A. CN V-1. Ophthalmic branch.

B. FRONTAL NERVE.

C. SUPRAORBITAL NERVE.

D. LACRIMAL NERVE.

E. NASOCILIARY NERVE.

F. POSTERIOR ETHMOID NERVE.

G. ANTERIOR ETHMOID NERVE.

H. INFRATROCHLEAR NERVE.

I. CN III. Oculomotor nerve.

J. CILIARY GANGLION.

K. CN IV. Trochlear nerve.

L. CN VI. Abducens nerve.

M. CN II. Optic nerve.

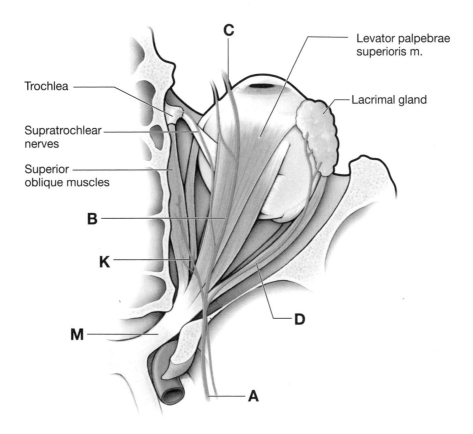

Levator palpebrae superioris m.

Lacrimal gland

Trochlea

Supratrochlear nerves

Superior oblique muscles

C

B

K

M

D

A

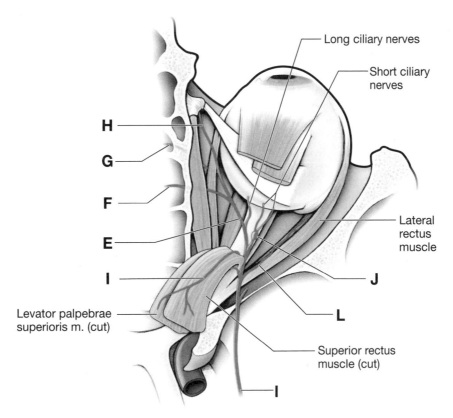

Long ciliary nerves

Short ciliary nerves

H

G

F

E

I

Levator palpebrae superioris m. (cut)

Lateral rectus muscle

J

L

Superior rectus muscle (cut)

I

Superior views of the nerves of the orbit (superficial and deep).

| 18f | **Arterial Supply of the Orbit** |

The ophthalmic artery arises from the internal carotid artery, courses through the optic canal and provides the principal vascular supply to the orbit and anterior region of the scalp.

A. INTERNAL CAROTID ARTERY.

B. OPHTHALMIC ARTERY.

C. CENTRAL RETINA ARTERY.

D. ETHMOIDAL ARTERIES.

E. SUPRATROCHLEAR ARTERY.

F. SUPRAORBITAL ARTERY.

G. LACRIMAL ARTERY.

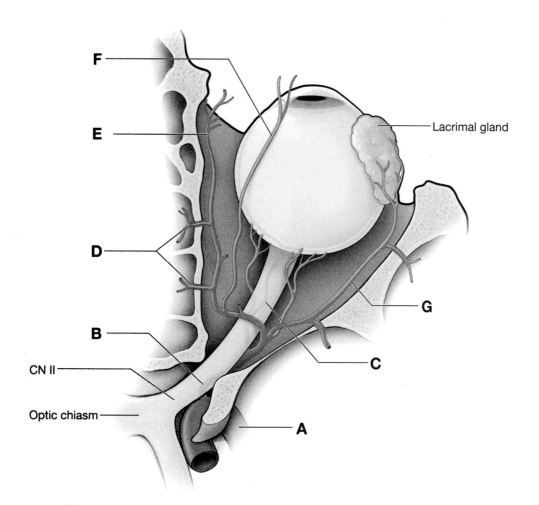

F

E

D

B

CN II

Optic chiasm

A

Lacrimal gland

G

C

Arteries of the orbit.

18g   Venous Drainage of the Orbit

The superior and inferior ophthalmic veins course through the superior and inferior orbital fissures and drain anteriorly to the facial vein, posteriorly to the cavernous sinus, and inferiorly to the pterygoid plexus.

A. CAVERNOUS SINUS.

B. SUPERIOR OPHTHALMIC VEIN.

C. SUPRAORBITAL VEIN.

D. PTERYGOID PLEXUS OF VEINS.

E. INFERIOR OPHTHALMIC VEIN.

F. INFRAORBITAL VEIN.

G. FACIAL VEIN.

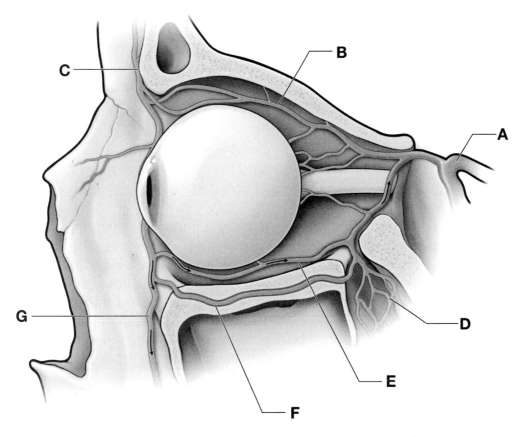

Veins of the orbit.

19.

....

Ear

19a	Ear

The external ear collects sound waves and transports them through the external acoustic meatus to the tympanic membrane. The tympanic membrane vibrates, setting three tiny ear ossicles (malleus, incus, and stapes) in the middle ear into motion. The stapes attaches to the lateral wall of the inner ear, where the vibration is transduced into fluid movement. The fluid causes the basilar membrane in the cochlea to vibrate. The vestibulocochlear nerve [cranial nerve (CN) VIII] receives and conducts the impulses to the brain, where there is integration of sound.

A. EXTERNAL ACOUSTIC MEATUS.

B. TYMPANIC MEMBRANE.

C. MALLEUS.

D. INCUS.

E. STAPES.

F. SEMICIRCULAR CANALS.

G. COCHLEA.

H. AUDITORY TUBE.

I. CN VII AND CN VIII.

J. SCALA VESTIBULI. Contains perilymph.

K. COCHLEAR DUCT. Contains endolymph.

L. TECTORIAL MEMBRANE.

M. SCALA TYMPANI. Contains perilymph.

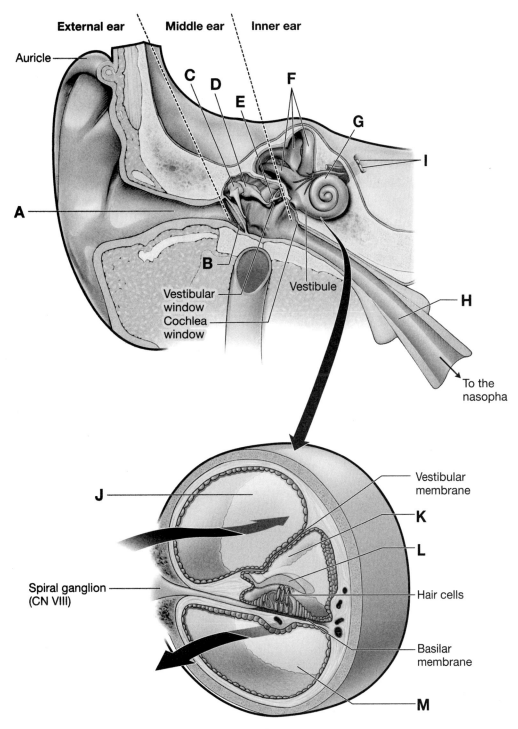

External ear **Middle ear** **Inner ear**

Auricle

C
D
E
F
G
I
A
B

Vestibular
window
Cochlea
window

Vestibule

H

To the
nasopha

Vestibular
membrane
K
L

Spiral ganglion
(CN VIII)

Hair cells

J

Basilar
membrane

M

**Coronal section of the temporal bone showing the external, middle, and
internal ear; coronal section of the cochlea.**

20.

Superficial Face

20a | Superficial Nerves of the Face

The sensory innervation of the face is provided by the three divisions of the trigeminal nerve (CN V), with each division supplying the upper, middle, and lower part of the face. The facial artery and the superficial temporal artery provide vascular supply.

A. SUPRAORBITAL NERVE (CN V-1).

B. SUPRATROCHLEAR NERVE (CN V-1).

C. EXTERNAL NASAL NERVE (CN V-1).

D. INFRAORBITAL NERVE (CN V-2).

E. MENTAL NERVE (CN V-3).

F. FACIAL NERVE (CN VII).

G. GREAT AURICULAR NERVE (C2).

H. EXTERNAL CAROTID ARTERY.

I. SUPERFICIAL TEMPORAL ARTERY.

J. FACIAL ARTERY.

K. LABIAL ARTERIES.

L. LATERAL NASAL ARTERY.

M. EXTERNAL JUGULAR VEIN.

N. PAROTID GLAND.

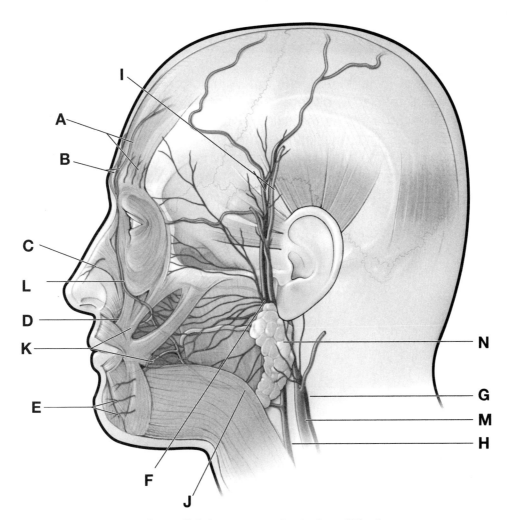

Superficial nerves and arteries of the face.

20b Muscles of Facial Expression

The muscles of facial expression are voluntary muscles located in the superficial fascia. In general, they arise from bones or fascia of the skull and insert into the skin, which enables a wide array of facial expression. The muscles are sometimes indistinct at their borders because they develop embryologically from a continuous sheet of musculature derived from the second branchial arch. The muscles of facial expression are located superficially in the neck, face, and scalp. Each muscle is innervated by CN VII (except the muscles of mastication, which are innervated by CN V-3).

A. FRONTALIS MUSCLE.

B. CORRUGATOR SUPERCILI MUSCLE.

C. ORBICULARIS OCULI MUSCLE.

D. PROCERUS MUSCLE.

E. LEVATOR LABII SUPERIORIS ALAEQUE NASI MUSCLE.

F. NASALIS MUSCLE.

G. ZYGOMATICUS MUSCLES.

H. LEVATOR LABII SUPERIORIS MUSCLE.

I. LEVATOR ANGULI ORIS MUSCLE.

J. RISORIUS MUSCLE.

K. PLATYSMA MUSCLE.

L. DEPRESSOR ANGULI ORIS MUSCLE.

M. DEPRESSOR LABII INFERIORIS MUSCLE.

N. MENTALIS MUSCLE.

O. ORBICULARIS ORIS MUSCLE.

P. BUCCINATOR MUSCLE.

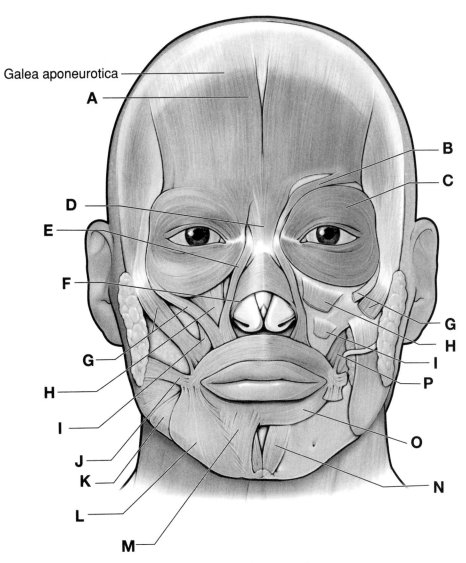

Muscles of facial expression.

20c   Facial Nerve Branches in the Face

CN VII provides motor innervation to the muscles of facial expression. The facial nerve exits the skull through the stylomastoid foramen and immediately gives off the posterior auricular nerve and other branches that supply the occipitalis, stylohyoid, and posterior digastricus muscles and the posterior auricular muscle. CN VII courses superficial to the external carotid artery and the retromandibular vein, enters the parotid gland, and divides into the following five terminal branches: temporal, zygomatic, buccal, mandibular, and cervical nerves, which in turn supply the muscles of facial expression. Other muscles of the face include muscles of mastication (temporalis, masseter, and the medial pterygoid and lateral pterygoid muscles), which are innervated by the motor division of CN V-3.

A. PAROTID GLAND.

B. FACIAL NERVE (CN VII).

C. POSTERIOR AURICULAR NERVE.

D. TEMPORAL BRANCHES.

E. ZYGOMATIC BRANCHES.

F. BUCCAL BRANCHES.

G. MANDIBULAR BRANCHES.

H. CERVICAL BRANCHES.

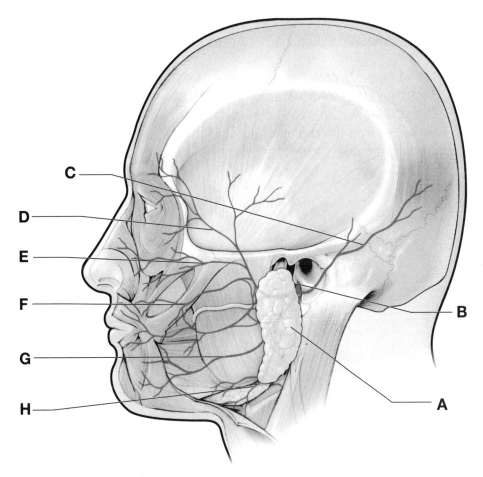

Branches of the facial nerve (CN VII) in the face.

20d   Parotid Gland

The parotid gland is situated in the lateral part of the face on the surface of the masseter muscle, anterior to the sternocleidomastoid muscle. A dense fascia covers the gland. The parotid gland produces and secretes saliva into the oral cavity and is innervated by visceral motor parasympathetic fibers from the glossopharyngeal nerve (CN IX).

A. PAROTID GLAND.

B. FACIAL NERVE (CN VII).

C. STERNOCLEIDOMASTOID MUSCLE.

D. EXTERNAL CAROTID ARTERY.

E. RETROMANDIBULAR VEIN.

F. SUPERIOR PHARYNGEAL CONSTRICTOR MUSCLE.

G. MASSETER MUSCLE.

H. BUCCINATOR MUSCLE.

I. INTERNAL JUGULAR VEIN.

J. INTERNAL CAROTID ARTERY.

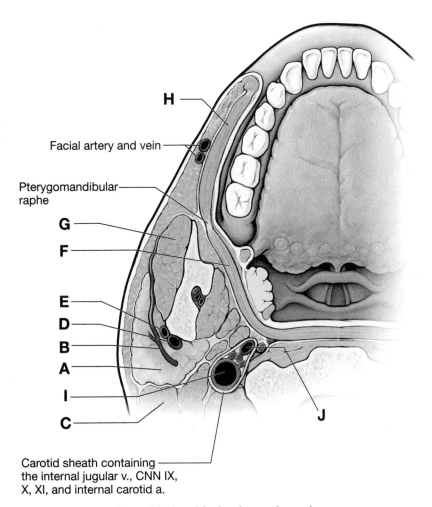

H

Facial artery and vein

Pterygomandibular
raphe

G

F

E

D

B

A

I

C

J

Carotid sheath containing
the internal jugular v., CNN IX,
X, XI, and internal carotid a.

Parotid gland in horizontal section.

21.

Infratemporal Fossa

21a   Muscles of Mastication

The muscles acting upon the temporomandibular joint (TMJ) are primarily the muscles that generate the various movements associated with chewing; hence, these muscles are often called the muscles of mastication. The branchial motor division of CN V-3 innervates each of the following muscles:

A. TEMPORALIS MUSCLE.

B. MASSETER MUSCLE.

C. LATERAL PTERYGOID MUSCLE.

D. MEDIAL PTERYGOID MUSCLE.

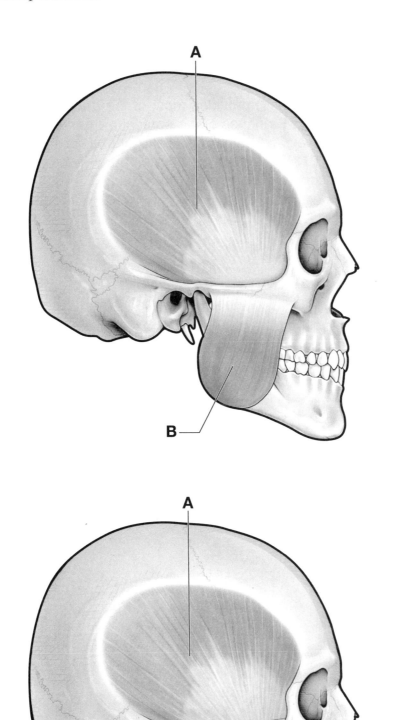

Muscles of mastication (superficial and deep views).

21b   Nerves of the Infratemporal Fossa

The trigeminal nerve (CN V) provides most of the general sensory innervation to the head. Branches of the ophthalmic nerve (CN V-1) and the maxillary nerve (CN V-2) provide only general sensory innervation. In contrast, CN V-3 enters the infratemporal fossa and has both a general sensory and a branchial motor root.

A. MANDIBULAR NERVE (CN V-3).

B. AURICULOTEMPORAL NERVE.

C. INFERIOR ALVEOLAR NERVE.

D. NERVE TO THE MYLOHYOID.

E. LONG BUCCAL NERVE.

F. LINGUAL NERVE.

G. MENTAL NERVE.

H. SUBMANDIBULAR GANGLION.

I. CHORDA TYMPANI NERVE. Branch of facial nerve (CN VII) that joins with the lingual nerve (CN V-3).

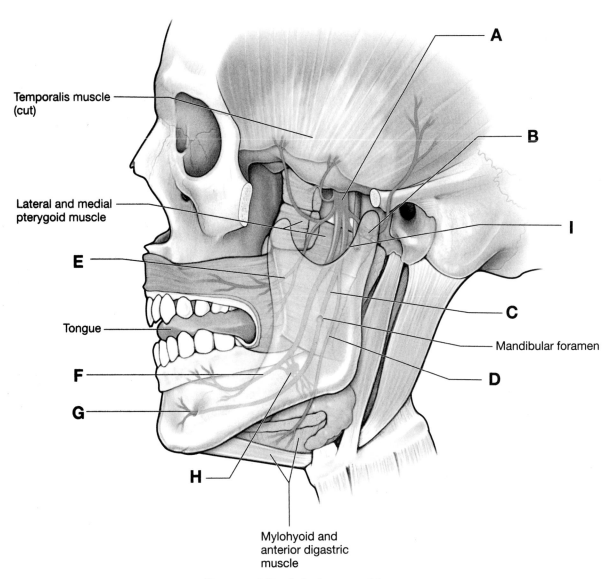

Temporalis muscle
(cut)

Lateral and medial
pterygoid muscle

E

Tongue

F

G

A

B

I

Mandibular foramen

C

D

H

Mylohyoid and
anterior digastric
muscle

Nerves of the infratemporal fossa.

22.

Pterygopalatine Fossa

22a Pterygopalatine Fossa Boundaries

The pterygopalatine fossa is an irregular space where neurovascular structures course through to the nasal cavity, palate, pharynx, orbit, and face. The neurovascular structures enter and exit the fossa through the following boundaries:

A. PTERYGOMAXILLARY FISSURE.

B. INFRAORBITAL GROOVE.

C. SPHENOPALATINE FORAMEN.

D. FORAMEN ROTUNDUM.

E. PHARYNGEAL CANAL.

F. PTERYGOID CANAL.

G. LESSER PALATINE CANAL.

H. GREATER PALATINE CANAL.

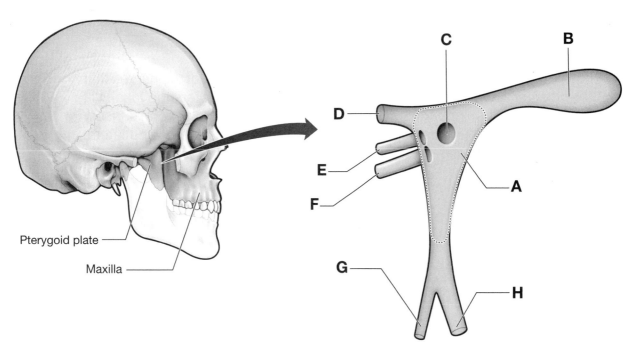

Pterygoid plate

Maxilla

Boundaries of the pterygopalatine fossa.

22b   Nerves of the Pterygopalatine Fossa

Branches of the maxillary nerve (CN V-2) form most of the nerves that enter and exit the pterygopalatine fossa. CN V-2 enters the fossa via the foramen rotundum. The maxillary nerve provides general sensory innervation to the maxillary teeth, upper lip, lower eyelid, and lateral side of the nose.

A. MAXILLARY NERVE (CN V-2).

B. INFRAORBITAL NERVE.

C. SUPERIOR ALVEOLAR NERVES.

D. ZYGOMATIC NERVE.

E. PHARYNGEAL NERVE.

F. NERVE OF PTERYGOID CANAL.

G. LESSER PALATINE NERVE.

H. GREATER PALATINE NERVE.

I. PTERYGOPALATINE GANGLION.

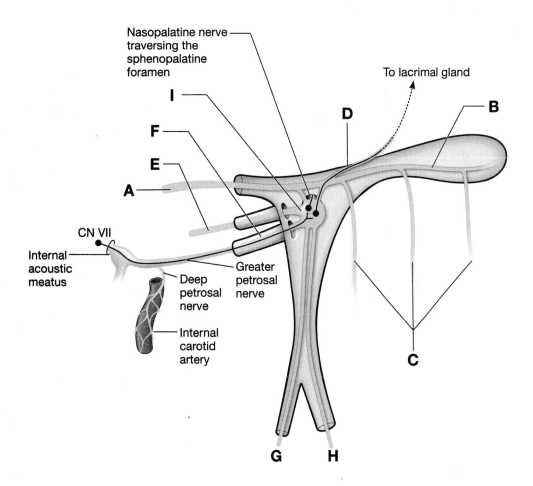

Nasopalatine nerve traversing the sphenopalatine foramen

To lacrimal gland

I

F

E

A

D

B

CN VII

Internal acoustic meatus

Deep petrosal nerve

Greater petrosal nerve

Internal carotid artery

C

G   H

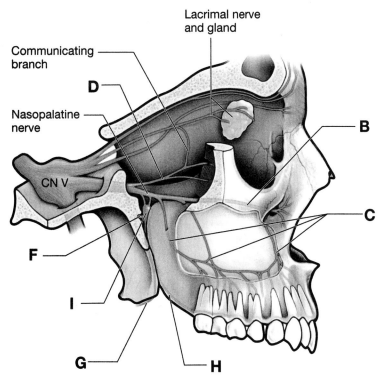

Lacrimal nerve and gland

Communicating branch

Nasopalatine nerve

D

CN V

B

F

C

I

G   H

Nerves of the pterygopalatine fossa.

23.

Nasal Cavity

23a   Nasal Cavity Boundaries

The nasal cavity is divided into two lateral compartments separated down the middle by the nasal septum. The nasal cavity communicates anteriorly through the nostrils and posteriorly with the nasopharynx through openings called choanae. The nasal cavities and septum are lined with a mucous membrane and are richly vascularized by branches of the maxillary, facial, and ophthalmic arteries.

A. NASAL SEPTAL CARTILAGE.

B. PERPENDICULAR PLANE OF ETHMOID BONE.

C. VOMER BONE.

D. PALATINE BONE.

E. MAXILLA.

F. NASAL BONE.

G. FRONTAL SINUS.

H. SUPERIOR NASAL CONCHA.

I. MIDDLE NASAL CONCHA.

J. INFERIOR NASAL CONCHA.

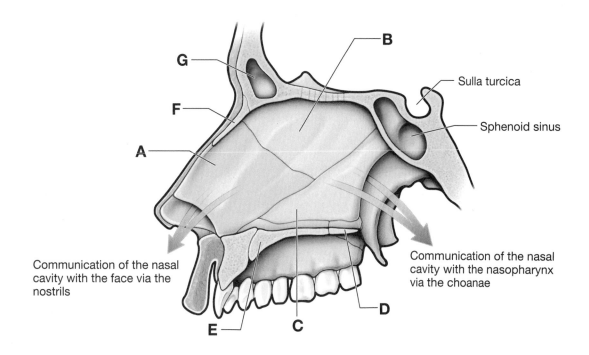

Sulla turcica

Sphenoid sinus

Communication of the nasal cavity with the face via the nostrils

Communication of the nasal cavity with the nasopharynx via the choanae

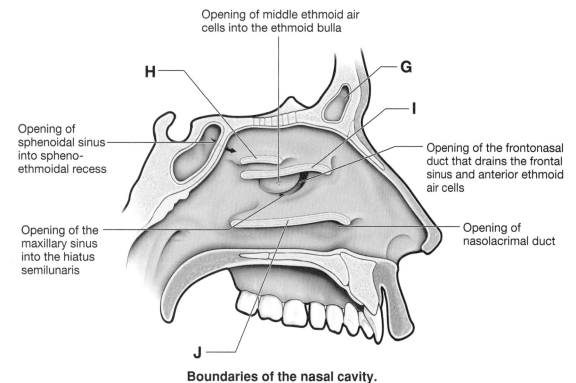

Opening of middle ethmoid air cells into the ethmoid bulla

Opening of sphenoidal sinus into spheno-ethmoidal recess

Opening of the maxillary sinus into the hiatus semilunaris

Opening of the frontonasal duct that drains the frontal sinus and anterior ethmoid air cells

Opening of nasolacrimal duct

Boundaries of the nasal cavity.

23b   Nerves of the Nasal Cavity

The superior region of the nasal cavity contains the olfactory nerves for smell. The remainder of the nerves are branches from the ophthalmic nerve (CN V-1) and maxillary nerve (CN V-2) for general sensory innervation.

A. OLFACTORY NERVES (CN I).

B. NASOPALATINE NERVE (SEPTAL BRANCH).

C. NASOPALATINE NERVE IN INCISIVE CANAL.

D. NASOPALATINE NERVE (LATERAL WALL BRANCH).

E. GREATER AND LESSER PALATINE NERVES.

F. ANTERIOR ETHMOID NERVE.

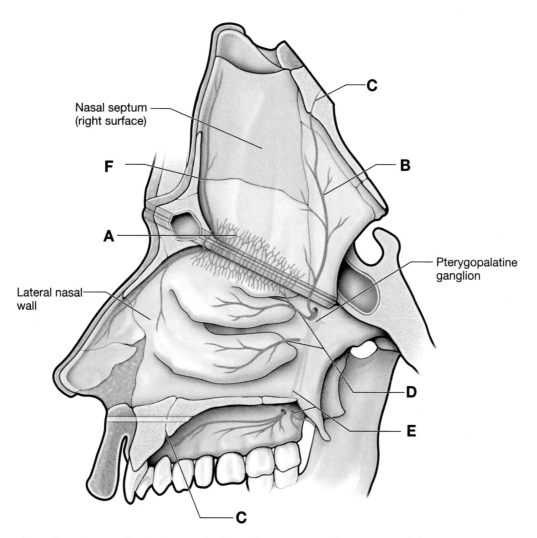

Nasal septum
(right surface)

C

F

B

A

Pterygopalatine
ganglion

Lateral nasal
wall

D

E

C

Nasal septum reflected superiorly to demonstrate the nerves of the nasal cavity.

| 23c | **Arteries of the Nasal Cavity** |

The nasal cavity receives its arterial supply from a number of arteries that form rich anastomotic connections.

A. EXTERNAL CAROTID ARTERY.

B. MAXILLARY ARTERY.

C. GREATER AND LESSER PALATINE ARTERIES.

D. SPHENOPALATINE ARTERY.

E. FACIAL ARTERY BRANCHES.

F. ETHMOIDAL ARTERY BRANCHES.

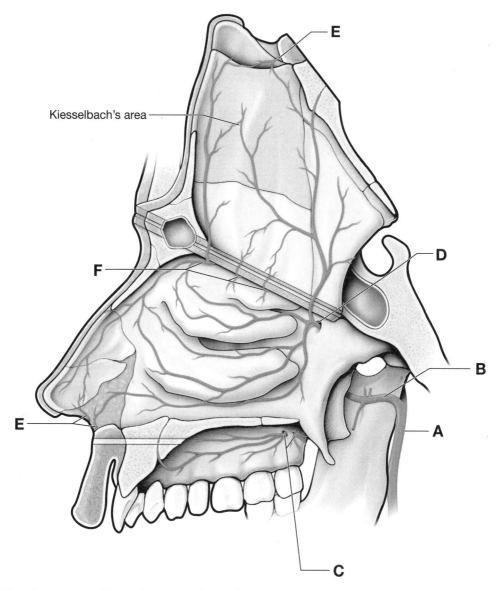

Kieselbach's area

Nasal septum reflected superiorly to demonstrate the arteries of the nasal cavity.

23d   Paranasal Sinuses

The paranasal sinuses are hollow cavities within the ethmoid, frontal, maxillary, and sphenoid bones. They help to decrease the weight of the skull, resonate sound produced through speech, and produce mucus. The paranasal cavities communicate with the nasal cavity, where mucus is drained. Branches of CN V provide general sensory innervation.

A. SPHENOID SINUS.

B. FRONTAL SINUS.

C. ETHMOID SINUSES.

D. MAXILLARY SINUS.

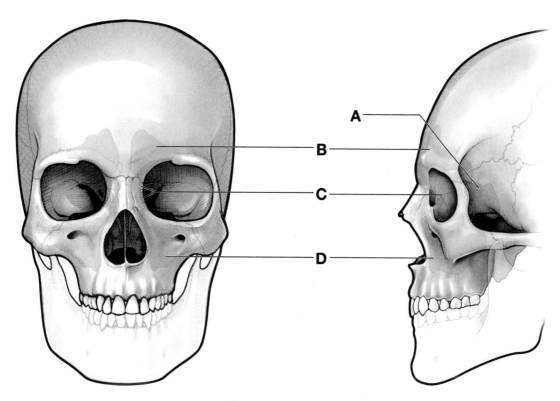

Paranasal sinuses.

24.

Oral Cavity

| 24a | **Sensory Innervation of the Palate** |

The palate forms both the roof of the oral cavity and the floor of the
nasal cavity and consists of a hard and a soft palate. Branches from
the maxillary nerve (CN V-2) provide general sensory innervation of
the palate.

A. GREATER PALATINE NERVE AND DISTRIBUTION.

B. LESSER PALATINE NERVE AND DISTRIBUTION.

C. NASOPALATINE NERVE AND DISTRIBUTION.

D. ANTERIOR SUPERIOR ALVEOLAR NERVE.

E. MIDDLE SUPERIOR ALVEOLAR NERVE.

F. POSTERIOR SUPERIOR ALVEOLAR NERVE.

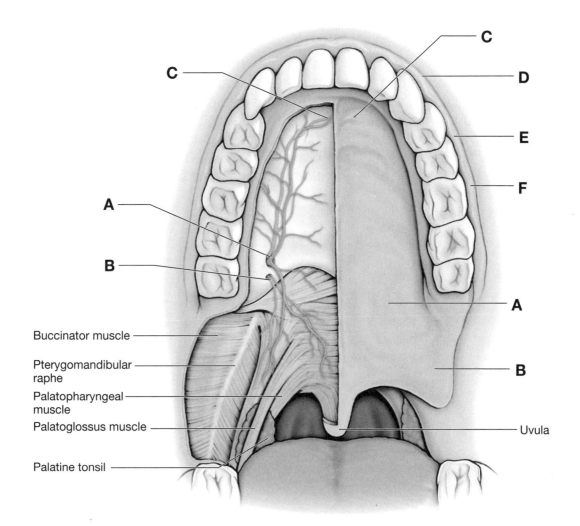

Buccinator muscle

Pterygomandibular raphe

Palatopharyngeal muscle

Palatoglossus muscle

Palatine tonsil

Uvula

Sensory innervation of the palate.

24b   Muscles of the Palate

The palate forms both the roof of the oral cavity and the floor of the nasal cavity and consists of a hard and a soft palate. All the muscles that act upon the soft palate are innervated by the vagus nerve [cranial nerve (CN) X], with the exception of the tensor veli palatini, which is innervated by a small motor branch from CN V-3. The difference in innervation reflects the embryologic origins of the branchial arches.

A. AUDITORY TUBE.

B. LEVATOR VELI PALATINE MUSCLE.

C. TENSOR VELI PALATINI MUSCLE.

D. PALATOPHARYNGEAL MUSCLE.

E. SUPERIOR PHARYNGEAL CONSTRICTOR MUSCLE.

F. MUSCULARIS UVULAE MUSCLE.

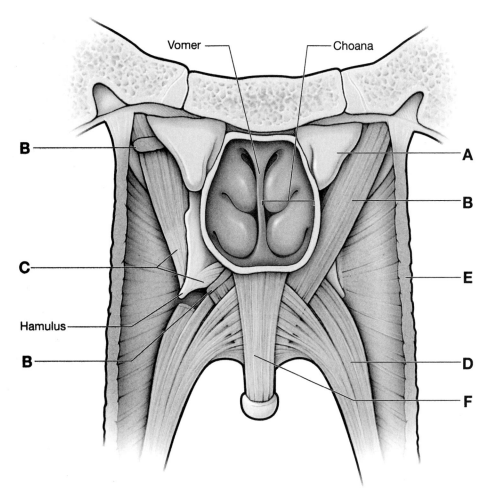

Vomer

Choana

B

C

Hamulus

B

A

B

E

D

F

**Posterior view of the palate with the mucosal lining removed
revealing the muscles.**

| 24c | **Muscles of the Tongue** |

The tongue consists of skeletal muscle, which has a surface covered with taste buds (special sensory) and general sensory nerve endings. Tongue muscles are bilaterally paired and innervated by the hypoglossal nerve (CN XII).

A. TONGUE.

B. PALATOGLOSSUS MUSCLE.

C. STYLOGLOSSUS MUSCLE.

D. STYLOHYOID MUSCLE.

E. HYOGLOSSUS MUSCLE.

F. GENIOGLOSSUS MUSCLE.

G. GENIOHYOID MUSCLE.

H. MYLOHYOID MUSCLE.

I. STYLOID PROCESS.

J. HYOID BONE.

K. MANDIBLE.

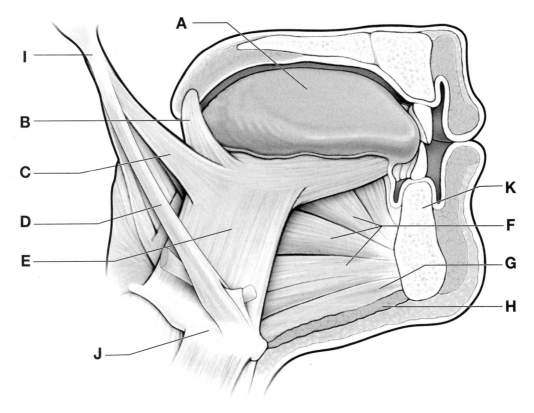

Sagittal section through oral cavity revealing muscles of the tongue.

24d   Neurovascular Supply of the Tongue

The tongue consists of skeletal muscle, which has a surface covered with taste buds (special sensory) and general sensory nerve endings. The tongue receives innervation from the lingual nerve (general sensation to anterior tongue), chorda tympani nerve (taste to anterior tongue), glossopharyngeal nerve (general sensation and taste to posterior tongue), and hypoglossal nerve (motor innervation of muscles). The lingual artery arises from the external carotid artery at the level of the tip of the greater horn of the hyoid bone in the carotid triangle. The lingual artery courses anteriorly between the hyoglossus and genioglossus muscles, giving rise to the dorsal lingual and sublingual branches and terminating as the deep lingual artery.

A. POSTERIOR THIRD OF TONGUE.

B. ANTERIOR THIRD OF TONGUE.

C. LINGUAL NERVE (CN V-3).

D. SUBMANDIBULAR GANGLION.

E. CHORDA TYMPANI NERVE (CN VII).

F. GLOSSOPHARYNGEAL NERVE (CN IX).

G. HYPOGLOSSAL NERVE (CN XII).

H. EXTERNAL CAROTID ARTERY.

I. DORSAL LINGUAL ARTERY.

J. LINGUAL ARTERY.

K. SUBLINGUAL ARTERY.

L. INTERNAL JUGULAR VEIN.

M. SUBMANDIBULAR DUCT.

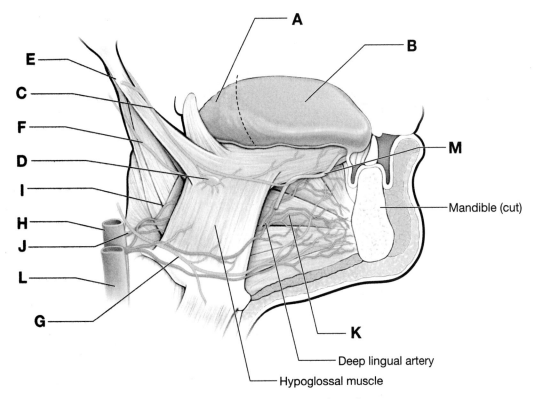

Sagittal section through oral cavity.

E
C
F
D
I
H
J
L
G
A
B
M

Mandible (cut)

Deep lingual artery

Hypoglossal muscle

24e   Teeth

The teeth cut, grind, and mix food during mastication. In adults, there are 16 teeth in the maxilla and 16 in the mandible. Branches of CN V-2 and the maxillary artery and veins supply the maxillary teeth and gingivae. Branches of CN V-3, and the inferior alveolar artery and veins supply the mandibular teeth and gingivae.

There are 20 teeth in a child and 32 in an adult. Adult teeth are typically numbered in a progressing clockwise fashion, with tooth number 1 (upper right maxillary molar) across to tooth number 16 (upper left maxillary molar). Tooth number 17 is the left third mandibular molar and continues to tooth number 32, the right third mandibular molar. The teeth are divided into four quadrants with eight teeth located in the upper left, upper right, lower left, and lower right halves of the maxilla and mandible. Each quadrant consists of 2 incisors, 1 canine, 2 premolars, and 3 molars.

A. INCISORS.

B. CANINE.

C. PREMOLARS.

D. MOLARS.

E. TRIGEMINAL NERVE (CN V).

F. MAXILLARY NERVE (CN V-2).

G. INFRAORBITAL NERVE.

H. POSTERIOR SUPERIOR ALVEOLAR NERVE.

I. MIDDLE SUPERIOR ALVEOLAR NERVE.

J. ANTERIOR SUPERIOR ALVEOLAR NERVE.

K. MANDIBULAR NERVE (CN V-3).

L. INFERIOR ALVEOLAR NERVE.

M. MENTAL NERVE.

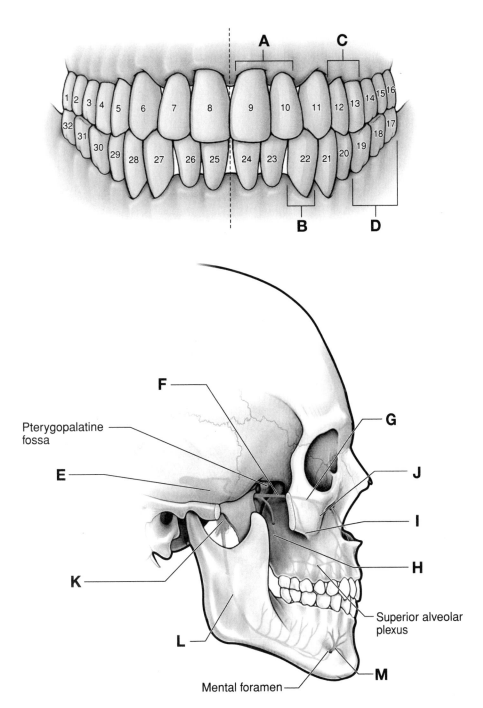

Structure and innervation of teeth.

SECTION V

NECK

25.

···

Overview of the Neck

25a   Fascial Layers of the Neck

The cervical fascia consists of concentric layers of fascia that compartmentalize structures in the neck. These fascial layers are defined as the superficial fascia and the deep fascia, with sublayers within the deep fascia. The sublayers include deep investing fascia, pretracheal fascia, prevertebral fascia, and the carotid sheath.

A. SKIN.

B. SUPERFICIAL FASCIA.

C. INVESTING FASCIA.

D. PRETRACHEAL FASCIA.

E. CAROTID SHEATH.

F. PREVERTEBRAL FASCIA.

Key for structures contained within the cervical fascia:

1. Thyroid gland

2. Infrahyoid muscles

3. Sternocleidomastoid muscle

4. Common carotid artery

5. Internal jugular vein

6. Vagus nerve

7. Sympathetic trunk

8. Platysma muscle

9. Anterior scalene muscle

10. Middle scalene muscle

11. Posterior scalene muscle

12. Trachea

13. Esophagus

14. Trapezius muscle

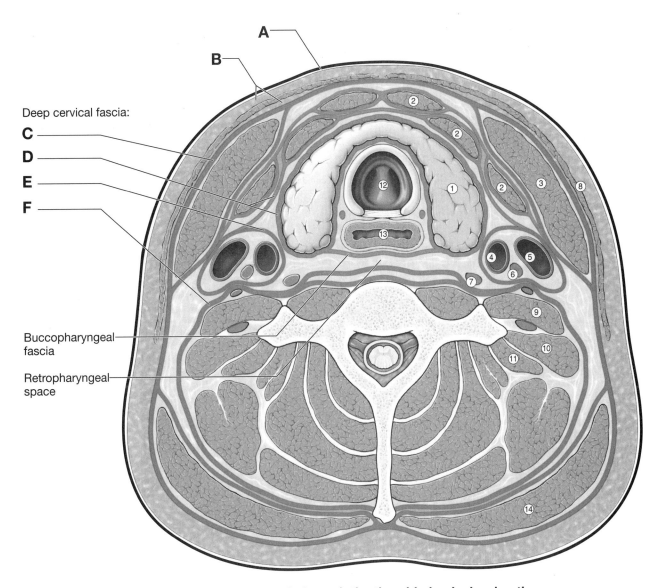

A

B

Deep cervical fascia:

C

D

E

F

Buccopharyngeal fascia

Retropharyngeal space

Cross-section of the neck through the thyroid gland, showing the layers of the cervical fascia.

25b   Neck Muscles

The muscles of the neck are organized and grouped with the cervical fascia. The platysma muscle is located within the superficial fascia, and the sternocleidomastoid and trapezius muscles are located within the investing fascia (part of the deep cervical fascia). Vertebral muscles (prevertebral, scalene, and deep cervical) are located within the prevertebral fascia. The suprahyoid muscles are deep to the investing fascia, whereas the infrahyoid muscles are within the pretracheal fascia.

A. STERNOCLEIDOMASTOID MUSCLE.

B. TRAPEZIUS MUSCLE.

C. PLATYSMAS MUSCLE.

D. ANTERIOR SCALENE MUSCLE.

E. MIDDLE SCALENE MUSCLE.

F. POSTERIOR SCALENE MUSCLE.

G. LONGUS CAPITIS MUSCLE.

H. LONGUS COLI MUSCLE.

I. OMOHYOID MUSCLE.

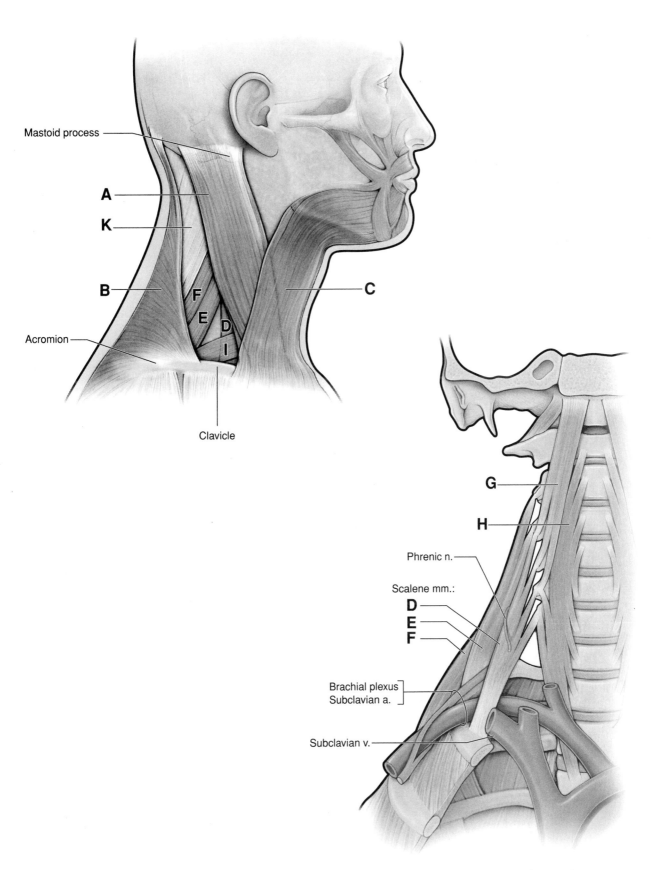

Mastoid process

A

K

B

Acromion

F

E

D

I

Clavicle

C

G

H

Phrenic n.

Scalene mm.:

D
E
F

Brachial plexus
Subclavian a.

Subclavian v.

A. Muscles of the neck. B. Anterior view of the scalene and prevertebral muscles.

25c   Hyoid Muscles

The suprahyoid muscles (i.e., digastric, stylohyoid, mylohyoid, and geniohyoid muscles) are located deep to the investing fascia of the deep cervical fascia. These muscles raise the hyoid bone during swallowing because the mandible is stabilized.

The infrahyoid muscles include four pairs of muscles that are located within the muscular layer of the pretracheal fascia, inferior to the hyoid bone (hence, the name). Each muscle is innervated by the ansa cervicalis from the cervical plexus (ventral rami C1–C3). Collectively, these muscles function to depress the hyoid bone and larynx during swallowing and speaking. They receive their names according to their attachments.

A. HYOID BONE.

B. DIGASTRIC MUSCLE.

C. STYLOHYOID MUSCLE.

D. MYLOHYOID MUSCLE.

E. HYOGLOSSUS MUSCLE.

F. OMOHYOID MUSCLE.

G. STERNOHYOID MUSCLE.

H. THYROHYOID MUSCLE.

I. STERNOTHYROID MUSCLE.

J. THYROID CARTILAGE.

K. THYROID GLAND.

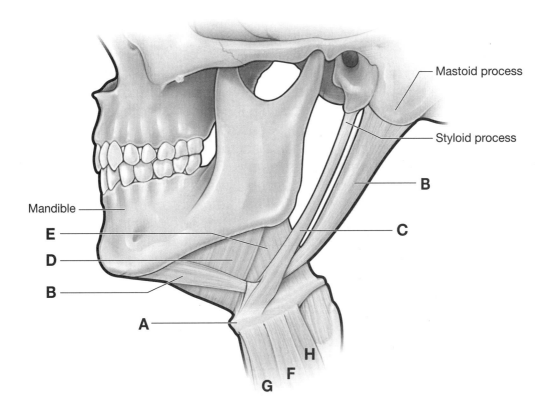

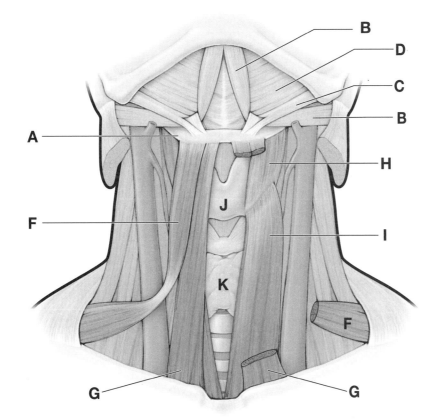

**Lateral view of the floor of the mouth, highlighting the suprahyoid muscles
(the geniohyoid muscle is not shown, top); anterior view of a step dissection,
highlighting the infrahyoid muscles (bottom).**

25d   Arteries of the Neck

The common carotid artery branches from the brachiocephalic artery on the right side and directly from the aortic arch on the left side. The common carotid artery ascends within the carotid sheath, along with the internal jugular vein and the vagus nerve. The common carotid artery bifurcates at the upper border of the thyroid cartilage into an internal and an external carotid artery. The internal carotid artery gives off no branches in the neck, but within the skull it provides vascular supply to the anterior and middle regions of the brain, the orbit and the scalp. The external carotid artery gives rise to numerous branches that supply the neck and face.

A. BRACHIOCEPHALIC ARTERY.

B. RIGHT SUBCLAVIAN ARTERY.

C. VERTEBRAL ARTERY.

D. THYROCERVICAL TRUNK.

E. COMMON CAROTID ARTERY.

F. INTERNAL CAROTID ARTERY.

G. EXTERNAL CAROTID ARTERY.

H. SUPERIOR LARYNGEAL ARTERY.

I. SUPERIOR THYROID ARTERY.

J. ASCENDING PHARYNGEAL ARTERY.

K. LINGUAL ARTERY.

L. FACIAL ARTERY.

M. OCCIPITAL ARTERY.

N. POSTERIOR AURICULAR ARTERY.

O. MAXILLARY ARTERY.

P. SUPERFICIAL TEMPORAL ARTERY.

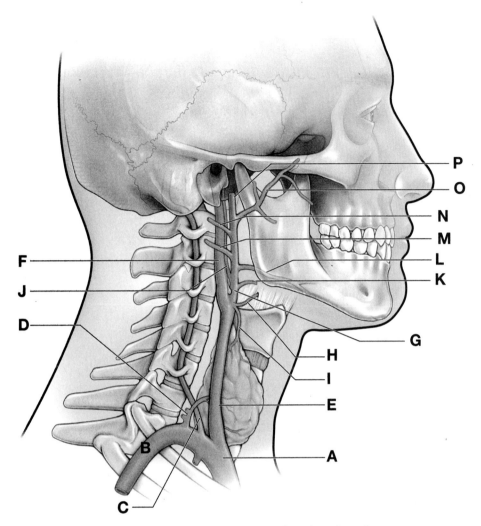

The principal arteries of the head and neck.

25e	**Veins of the Neck**

The external and anterior jugular veins are the principal venous return for the neck, and the internal jugular vein provides venous return for the head.

The external jugular vein is forward at the angle of the mandible via the joining of the posterior auricular and posterior branch of the retro-mandibular veins. The external jugular vein descends vertically down the neck within the superficial fascia, deep to the platysma muscle. After crossing the sternocleidomastoid muscle, the external jugular vein pierces the deep investing fascia posterior to the clavicular head and enters the subclavian vein.

The internal jugular vein originates at the jugular foramen by the union of the sigmoid and inferior petrosal sinuses and serves as the principal drainage of the skull, brain, superficial face, and parts of the neck. After exiting the skull via the jugular foramen, along with the glossopharyngeal, vagus, and accessory nerves (CNN IX, X, and XI, respectively), the internal jugular vein traverses the neck within the carotid sheath. The internal jugular vein joins with the subclavian vein to form the brachiocephalic vein.

A. INTERNAL JUGULAR VEIN.

B. EXTERNAL JUGULAR VEIN.

C. RETROMANDIBULAR VEIN.

D. POSTERIOR AURICULAR VEIN.

E. PTERYGOID PLEXUS OF VEINS.

F. MAXILLARY VEIN.

G. SUPERFICIAL TEMPORAL VEIN.

H. FACIAL VEIN.

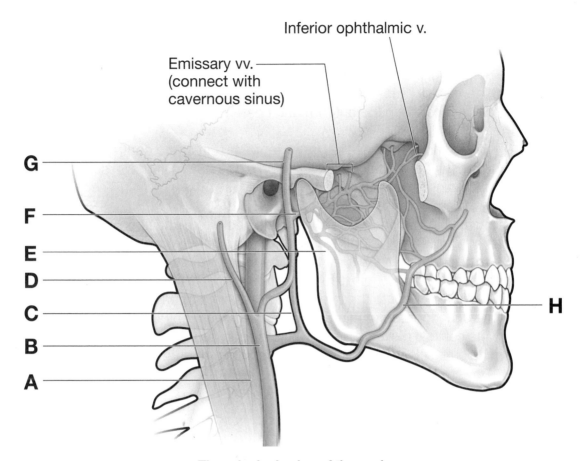

Emissary vv.
(connect with
cavernous sinus)

Inferior ophthalmic v.

G

F

E

D

C

B

A

H

The principal veins of the neck.

25f   Cervical Plexus of Nerves

The cervical plexus of nerves (sensory and motor components) arises from the ventral rami of cervical nerves C1 to C4 and exits the vertebral column between the anterior and posterior scalene muscles.

Sensory branches pierce the prevertebral fascia at the central region of the posterior border of the sternocleidomastoid muscle serving various regions of the skin of the neck.

Motor branches innervate muscles of the neck wall. The fibers from C1 travel briefly with CN XII and innervate the thyrohyoid and geniohyoid muscles. Fibers from C1 form the superior root of the ansa cervicalis. Fibers from C2 and C3 join to form the inferior root of the ansa cervicalis, where it lies anterior to the internal jugular vein and passes upward to join the superior root, forming a loop (ansa). Most of the motor nerves from the cervical plexus branch from the ansa cervicalis, supplying the infrahyoid muscles (sternothyroid, sternohyoid, and omohyoid).

A. LESSER OCCIPITAL NERVE.

B. GREAT AURICULAR NERVE.

C. TRANSVERSE CERVICAL NERVE.

D. SUPRACLAVICULAR NERVE.

E. PHRENIC NERVE.

F. SUPERIOR ROOT OF ANSA CERVICALIS.

G. INFERIOR ROOT OF ANSA CERVICALIS.

H. NERVE TO GENIOHYOID AND THYROHYOID.

I. NERVE TO SUPERIOR BELLY OF OMOHYOID.

J. NERVE TO STERNOTHYROID.

K. NERVE TO STERNOHYOID.

L. NERVE TO INFERIOR BELLY OF OMOHYOID.

M. HYPOGLOSSAL NERVE (CN XII).

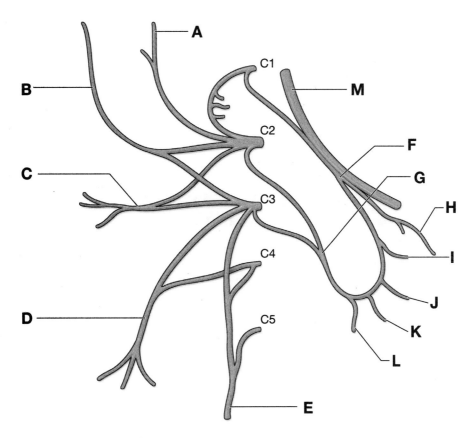

Schematic of the cervical plexus.

26.

Triangles and Root of the Neck

26a	Cervical Plexus of Nerves In Situ

The sensory branches of the cervical plexus course between the anterior and middle scalene muscles piercing the investing fascia at the posterior border of the sternocleidomastoid muscle en route to their respective cutaneous fields.

A. LESSER OCCIPITAL NERVE.

B. GREAT AURICULAR NERVE.

C. TRANSVERSE CERVICAL NERVE.

D. SUPRACLAVICULAR NERVE.

E. EXTERNAL JUGULAR VEIN.

F. DEEP INVESTING FASCIA.

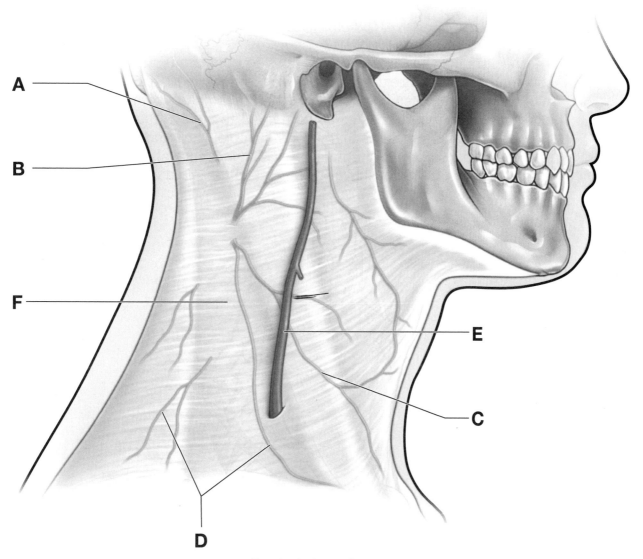

Cervical plexus in situ.

26b Triangles of the Neck

The neck region is divided into triangles to compartmentalize the contents. The sternocleidomastoid muscle divides the neck region into posterior and anterior triangles. The posterior triangle of the neck is located on the lateral aspect of the neck. The anterior triangle of the neck contains structures that enter the cranial vault and the thorax. The anterior triangle is further subdivided into smaller triangles: carotid triangle, submandibular triangle, submental triangle, and muscular triangle (not shown).

A. TRAPEZIUS MUSCLE.

B. STERNOCLEIDOMASTOID MUSCLE.

C. POSTERIOR TRIANGLE OF THE NECK.

D. ANTERIOR TRIANGLE OF THE NECK.

E. CAROTID TRIANGLE.

F. SUBMANDIBULAR TRIANGLE.

G. SUBMENTAL TRIANGLE.

H. DIGASTRIC MUSCLE.

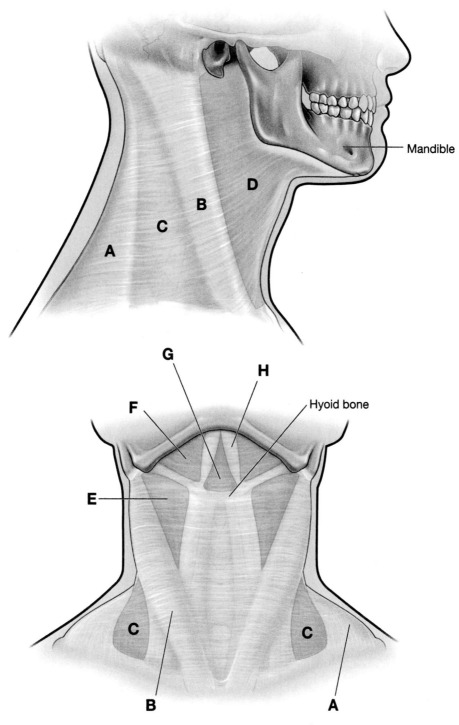

Triangles of the neck.

26c Thyroid and Parathyroid Glands

The thyroid gland regulates metabolic processes and decreases blood calcium concentration. The gland has two lateral lobes, which are connected by a central isthmus overlying the second through the fourth tracheal rings, all enclosed within the pretracheal fascia. The sternohyoid muscles cover each of the lateral lobes of the thyroid gland.

Parathyroid glands work antagonistically with the thyroid by increasing the blood calcium concentration. Usually, four parathyroid glands (two located on each side of the thyroid gland) are present on the deep surface of the thyroid gland.

A. THYROID GLAND.

B. TRACHEA.

C. THYROID CARTILAGE.

D. THYROHYOID MEMBRANE.

E. HYOID.

F. VAGUS NERVE (CN X).

G. VERTEBRAL ARTERY.

H. PHRENIC NERVE.

I. ESOPHAGUS.

J. AORTA.

K. SUPERIOR VENA CAVA.

L. BRACHIOCEPHALIC VEINS.

M. INTERNAL JUGULAR VEIN.

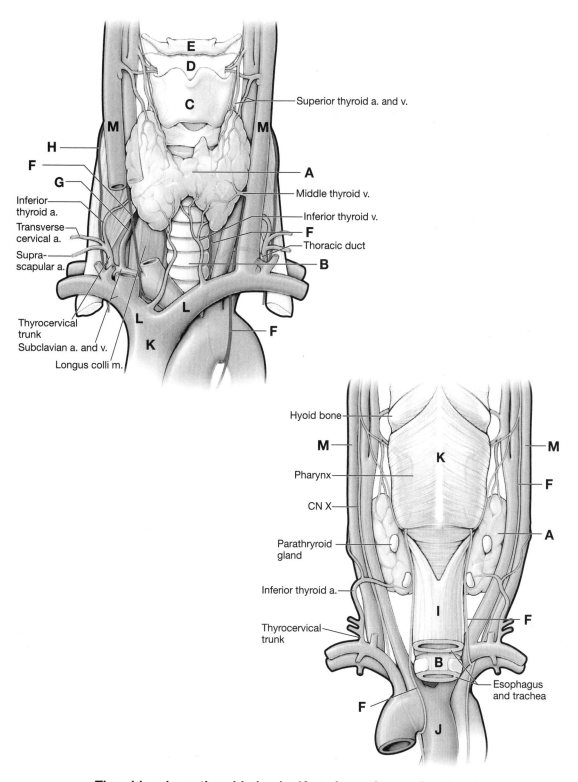

Superior thyroid a. and v.

Middle thyroid v.

Inferior thyroid v.

Thoracic duct

Inferior thyroid a.

Transverse cervical a.

Supra-scapular a.

Thyrocervical trunk

Subclavian a. and v.

Longus colli m.

Hyoid bone

Pharynx

CN X

Parathyroid gland

Inferior thyroid a.

Thyrocervical trunk

Esophagus and trachea

Thyroid and parathyroid glands. (Anterior and posterior views)

27.

Pharynx

27a   **Pharynx (Midsagittal Section)**

The pharynx is classically divided into three compartments, based on location: the nasopharynx, the oropharynx, and the laryngopharynx.

A. NASOPHARYNX.

B. OROPHARYNX.

C. LARYNGOPHARYNX.

D. ESOPHAGUS.

E. TONGUE.

F. NASAL SEPTUM.

G. CRICOID CARTILAGE.

H. THYROID CARTILAGE.

I. ORAL CAVITY.

J. HYOID.

K. SOFT PALATE.

L. TRACHEA.

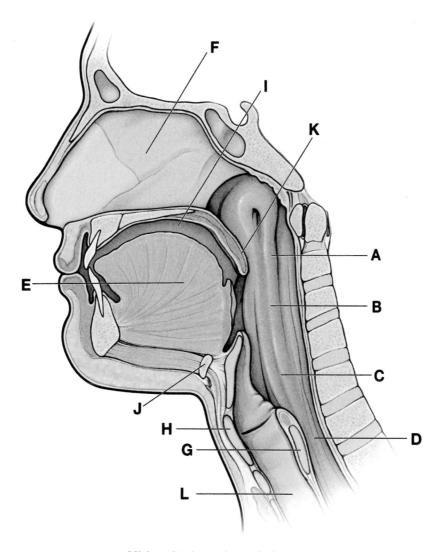

Midsagittal section of pharynx.

27b   Pharynx (Posterior View)

The constrictor muscles form the lateral and posterior walls of the pharynx and are attached posteriorly to the median pharyngeal raphe. The median pharyngeal raphe extends downward from the pharyngeal tubercle, on the base of the occipital bone anterior to the foramen magnum, and blends inferiorly with the posterior wall of the laryngopharynx and esophagus. The pharyngobasilar fascia separates the mucosa and the muscle layer, and blends with the periosteum of the base of the skull.

A. SUPERIOR PHARYNGEAL CONSTRICTOR.

B. MIDDLE PHARYNGEAL CONSTRICTOR.

C. INFERIOR PHARYNGEAL CONSTRICTOR.

D. ESOPHAGUS.

E. STYLOPHARYNGEUS MUSCLE.

F. DIGASTRIC MUSCLE.

G. STYLOHYOID MUSCLE.

H. THYROID GLAND.

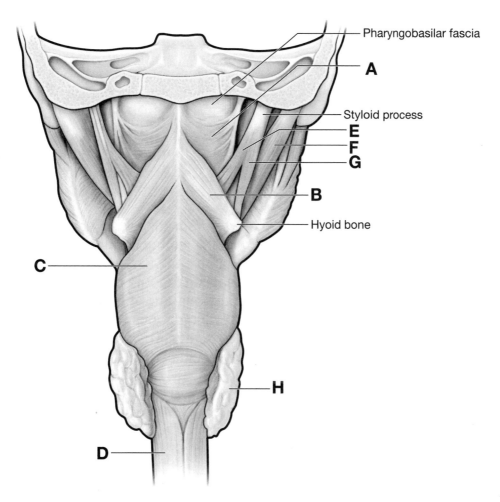

Pharyngobasilar fascia

A

Styloid process

E
F
G

B

Hyoid bone

C

H

D

Posterior view of pharynx.

28.

Larynx

28a	**Laryngeal Framework**

The larynx forms the air passageway from the hyoid bone to the trachea. The larynx is continuous with the laryngopharynx superiorly and with the trachea inferiorly. The larynx provides a patent (open) airway and acts as a switching mechanism to route air and food into the proper channels. The larynx is commonly known as the voice box and provides the cartilaginous framework for vocal fold and muscle attachments, which vibrate to produce sound. The cartilages include thyroid cartilage, cricoid cartilage, arytenoid cartilage, and epiglottis.

A. EPIGLOTTIS.

B. THYROID CARTILAGE.

C. CRICOID CARTILAGE.

D. TRACHEA.

E. HYOID BONE.

F. ARYTENOID CARTILAGE.

G. CORNICULATE CARTILAGE.

H. VOCAL LIGAMENTS.

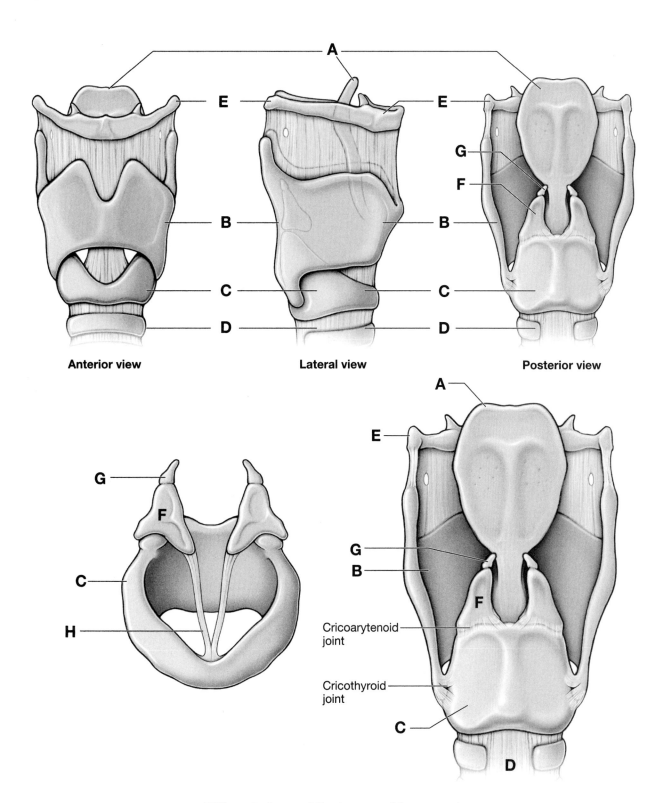

Anterior view   **Lateral view**   **Posterior view**

Cricoarytenoid
joint

Cricothyroid
joint

Different views of the laryngeal framework.

28b   Laryngeal Muscles

The intrinsic laryngeal muscles move the laryngeal framework, altering the size and shape of the rima glottidis and the length and tension of the vocal folds. The actions of the laryngeal muscles are best understood when considered in the following functional groups: adductors and abductors, and tensors and relaxers.

A. EPIGLOTTIS.

B. HYOID BONE.

C. THYROID CARTILAGE.

D. CORACOID CARTILAGE.

E. ARYTENOID CARTILAGE.

F. ARYTENOID MUSCLES.

G. POSTERIOR CRICOARYTENOIDEUS MUSCLE.

H. LATERAL CRICOARYTENOID MUSCLE.

I. THYROARYTENOID MUSCLE.

J. ARYEPIGLOTTICUS MUSCLE.

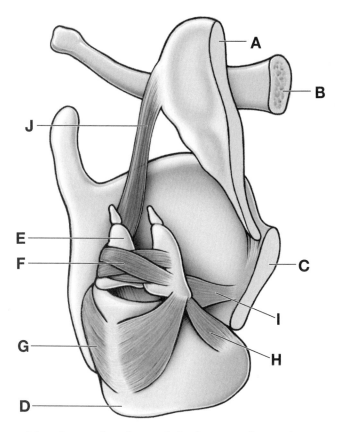

Muscles and actions of the laryngeal muscles.

SECTION VI

UPPER LIMB

29.

Overview of the Upper Limb

| 29a | Scapula |

The bones of the skeleton provide a framework to which soft tissues (e.g., muscles) can attach. The bony structure of the shoulder and arm, from proximal to distal, consists of the clavicle, scapula, and humerus. Synovial joints and ligaments connect bone to bone.

The scapula, or shoulder blade, is a large, flat triangular bone with three angles (lateral, superior, and inferior), three borders (superior, lateral, and medial), two surfaces (costal and posterior), and three processes (acromion, spine, and coracoid). The following landmarks are found on the scapula:

A. ACROMION.

B. CORACOID PROCESS.

C. GLENOID CAVITY.

D. LATERAL BORDER.

E. SUBSCAPULAR FOSSA.

F. SUPRASCAPULAR NOTCH.

G. SUPERIOR ANGLE.

H. SUPRASPINOUS FOSSA.

I. SPINE.

J. INFRASPINOUS FOSSA.

K. MEDIAL BORDER.

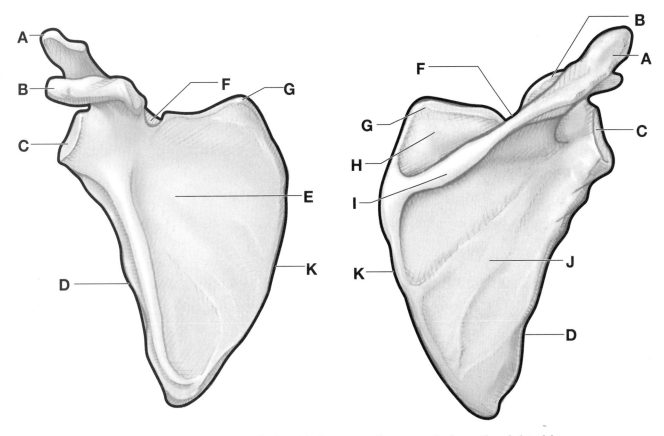

**Osteology of the upper limb with focus on the scapula from the right side
(anterior and posterior views).**

29b   Humerus

The humerus is the longest bone of the arm and is characterized by many distinct features that help to allow the upper extremity to move through a significant range of motion. The following landmarks are found on the humerus:

A. HEAD OF HUMERUS.

B. LESSER TUBERCLE.

C. GREATER TUBERCLE.

D. DELTOID TUBEROSITY.

E. LATERAL EPICONDYLE.

F. MEDIAL EPICONDYLE.

G. CORONOID FOSSA.

H. OLECRANON FOSSA.

I. CAPITULUM.

J. TROCHLEA.

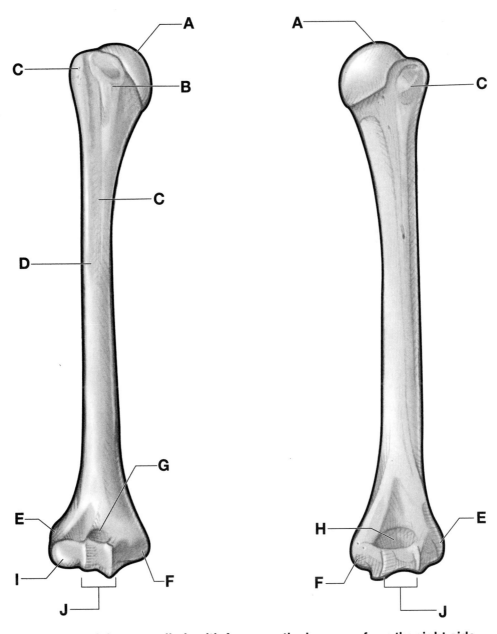

**Osteology of the upper limb with focus on the humerus from the right side
(left, anterior; right, posterior).**

29c   Radius, Ulna, and Hand

The bony structure of the forearm and hand consists of the radius, ulna, 8 carpals, 5 metacarpals, and 14 phalanges. The radius and ulna are bound together by a tough fibrous sheath known as the interosseous membrane.

In the anatomic position, the radius is the lateral bone of the forearm. It articulates with the capitulum of the humerus and the ulna. The ulna is the medial bone and articulates with the trochlea of the humerus and the radius.

The hand is subdivided into the carpus (wrist), metacarpus, and digits. The carpus is formed by eight small carpal bones arranged as a proximal row and a distal row, with each row consisting of four bones. Each of the five metacarpal bones is related to one digit. The first metacarpal is related to the thumb (digit 1), and metacarpals 2 through 5 are related to the index, middle, ring, and little finger, respectively. The phalanges are the bones of the five digits (numbered 1–5, beginning at the thumb). Digits 2 through 5 consist of a proximal, a middle, and a distal phalanx. Digit 1 (thumb) contains only a proximal and a distal phalanx.

A. RADIUS.

B. ULNA.

C. SCAPHOID BONE.

D. LUNATE BONE.

E. TRIQUETRUM BONE.

F. PISIFORM BONE.

G. TRAPEZIUM BONE.

H. TRAPEZOID BONE.

I. CAPITATE BONE.

J. HAMATE BONE.

K. METACARPAL BONES.

L. PROXIMAL PHALANGES.

M. MIDDLE PHALANGES.

N. DISTAL PHALANGES.

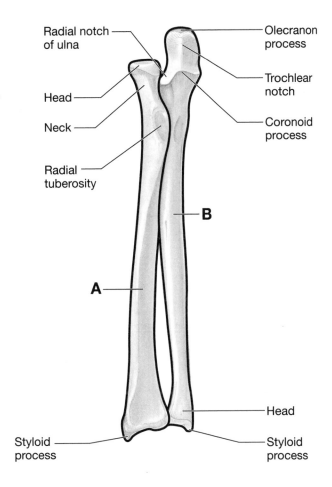

Radial notch of ulna

Olecranon process

Head

Trochlear notch

Neck

Coronoid process

Radial tuberosity

B

A

Head

Styloid process

Styloid process

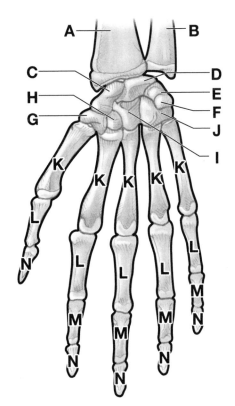

Osteology of the upper limb with focus on the radius, ulna, and hand from the right side (anterior view).

29d	**Compartments of the Arm and Forearm**

Two fascial layers, defined as the superficial and the deep fascia, lie between the skin and the bone of the upper limb. The deep fascia divides the upper limb into anterior and posterior compartments. Muscles are organized into these compartments and have common attachments, innervations, and actions.

A. ANTERIOR COMPARTMENT OF THE ARM. Muscles are innervated by the musculocutaneous nerve and primarily flex the elbow.

B. POSTERIOR COMPARTMENT OF THE ARM. Muscles are innervated by the radial nerve and primarily extend the elbow.

C. ANTERIOR COMPARTMENT OF THE FOREARM. Muscles are innervated by the median and ulnar nerves and primarily flex the wrist and digits.

D. POSTERIOR COMPARTMENT OF THE FOREARM. Muscles are innervated by the radial nerve and primarily extend the wrist and digits.

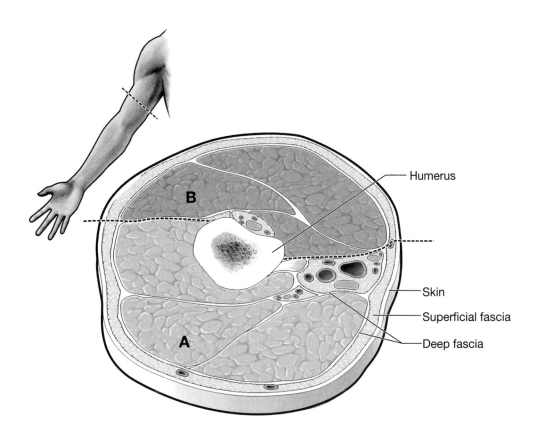

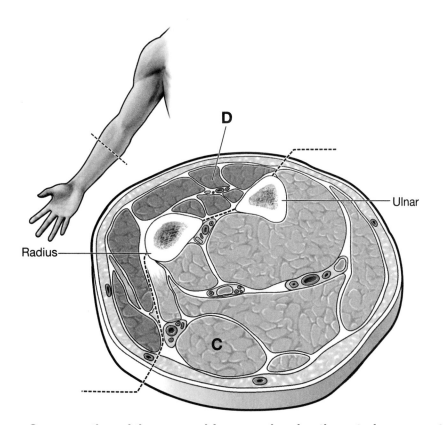

**Cross-section of the arm and forearm showing the anterior compartments
(flexors) and posterior compartments (extensors).**

29e   Brachial Plexus

The upper limb is innervated by anterior rami originating from spinal nerve levels C5–T1. These rami form a network of nerves referred to as the brachial plexus. The brachial plexus extends from the neck and courses distally through the axilla, providing motor and sensory innervation to the upper limb. As the brachial plexus courses distally, it forms roots, trunks, divisions, cords, and terminal branches.

A. ROOTS OF BRACHIAL PLEXUS.

B. UPPER TRUNK.

C. MIDDLE TRUNK.

D. LOWER TRUNK.

E. ANTERIOR DIVISION.

F. POSTERIOR DIVISION.

G. LATERAL CORD.

H. MEDIAL CORD.

I. POSTERIOR CORD.

J. MUSCULOCUTANEOUS NERVE.

K. MEDIAN NERVE.

L. ULNAR NERVE.

M. RADIAL NERVE.

N. AXILLARY NERVE.

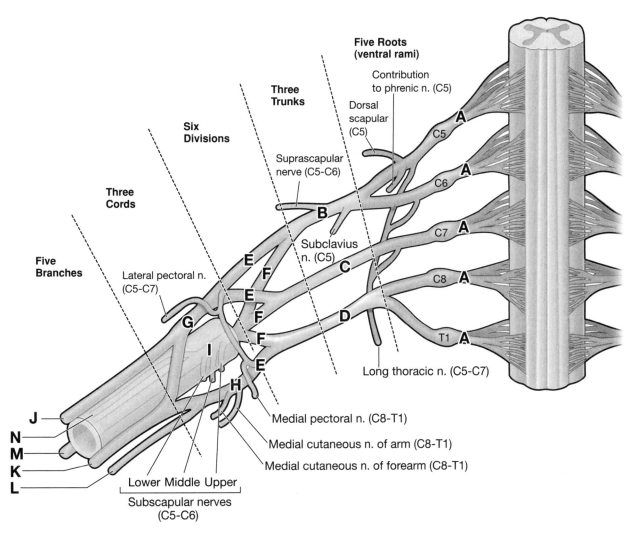

Five Roots (ventral rami)

Contribution to phrenic n. (C5)

Dorsal scapular (C5)

Three Trunks

Suprascapular nerve (C5-C6)

Six Divisions

Three Cords

Subclavius n. (C5)

Five Branches

Lateral pectoral n. (C5-C7)

Long thoracic n. (C5-C7)

Medial pectoral n. (C8-T1)

Medial cutaneous n. of arm (C8-T1)

Medial cutaneous n. of forearm (C8-T1)

Lower Middle Upper

Subscapular nerves (C5-C6)

Schematic of the brachial plexus showing the roots, trunks, divisions, cords, and terminal branches.

30.

Shoulder and Axilla

30a   Scapulothoracic Joint

The combined joints connecting the scapula (scapulothoracic joint), clavicle (sternoclavicular and acromioclavicular joints), and humerus (glenohumeral joint) form the shoulder complex and anchor the upper limb to the trunk. The only bony stability of the upper limb to the trunk is through the connection between the clavicle and the sternum. The remaining stability of the shoulder complex depends on muscles, and as a result, the shoulder complex has a wide range of motion.

The scapulothoracic joint is formed by the articulation of the scapula with the thoracic wall through the scapular muscles, including the trapezius and serratus anterior muscles.

The scapulothoracic joint is not considered a true anatomic joint; as such, it is frequently referred to as a "pseudo joint" because it does not contain the typical joint characteristics (e.g., synovial fluid and cartilage).

In addition, the scapulothoracic joint works in conjunction with the glenohumeral joint to produce movements of the shoulder. For example, the available range of motion for shoulder abduction is 180 degrees. This motion is produced by approximately 120 degrees from abduction at the glenohumeral joint and by approximately 60 degrees of upward rotation from the scapulothoracic joint.

A. CLAVICLE.

B. SCAPULA.

C. HEAD OF HUMERUS.

D. SUBSCAPULARIS MUSCLE.

E. SERRATUS ANTERIOR MUSCLE.

F. SCAPULOTHORACIC JOINT.

G. RHOMBOID MUSCLES.

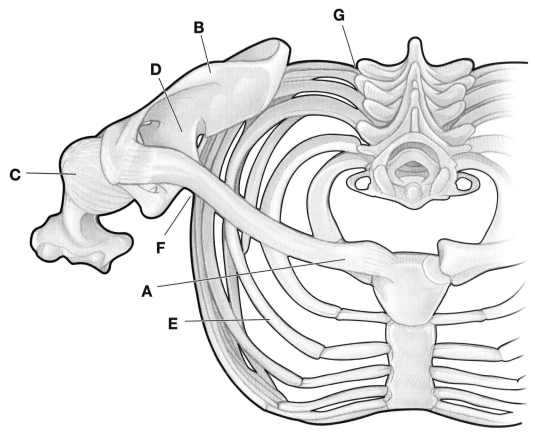

Superior view of the scapulothoracic joint.

30b — Movements of the Scapula and Glenohumeral Joint

The following illustrations demonstrate the movements of the scapula (scapulothoracic joint) and humerus (glenohumeral joint). The glenohumeral joint is a synovial, ball-and-socket joint. The "ball" is the head of the humerus, and the "socket" is the glenoid fossa of the scapula. The glenohumeral joint is considered to be the most mobile joint in the body.

A. ELEVATION OF SCAPULA.

B. DEPRESSION OF SCAPULA.

C. PROTRACTION OF SCAPULA (ABDUCTION).

D. RETRACTION OF SCAPULA (ADDUCTION).

E. UPWARD ROTATION.

F. DOWNWARD ROTATION.

G. FLEXION OF HUMERUS.

H. EXTENSION OF HUMERUS.

I. ABDUCTION OF HUMERUS.

J. ADDUCTION OF HUMERUS.

K. MEDIAL ROTATION.

L. LATERAL ROTATION.

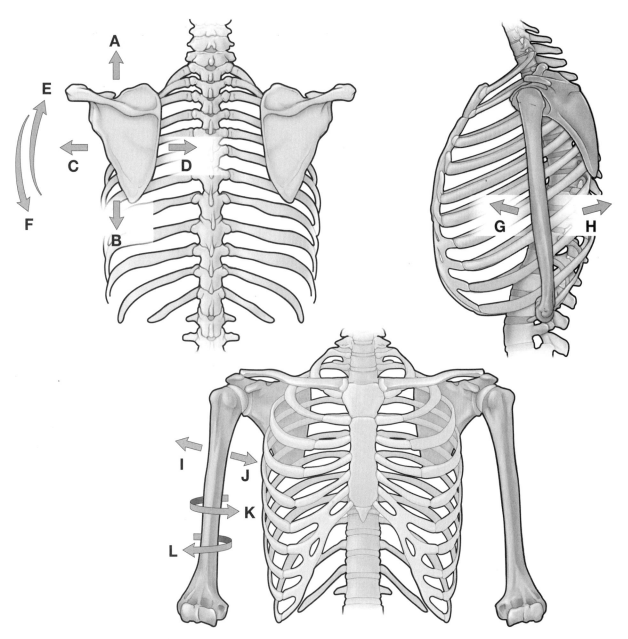

Movements of the scapulothoracic and glenohumeral joints.

30c Scapulothoracic Joint Muscles

The following muscles and muscle groups comprise the muscles of the scapulothoracic joint:

A. TRAPEZIUS MUSCLE.

B. LEVATOR SCAPULAE MUSCLE.

C. RHOMBOID MINOR MUSCLE.

D. RHOMBOID MAJOR MUSCLE.

E. SERRATUS ANTERIOR MUSCLE.

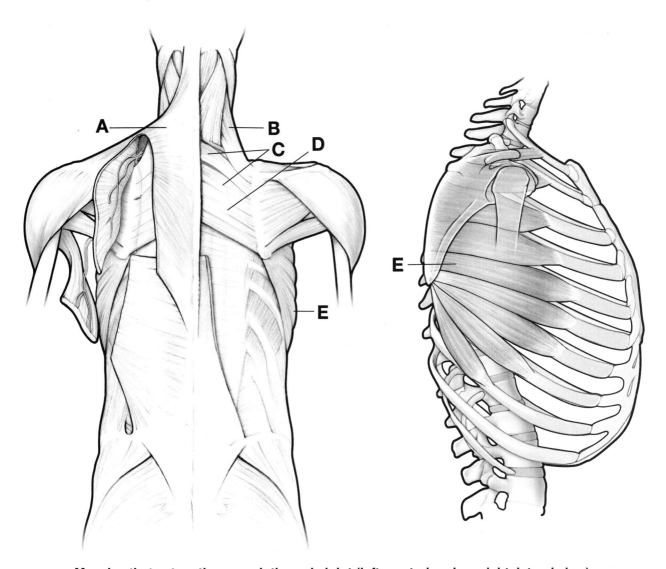

Muscles that act on the scapulothoracic joint (left, posterior view; right, lateral view).

| 30d | **Glenohumeral Joint Muscles** |

The following muscles and muscle groups comprise the muscles of the glenohumeral joint:

A. DELTOID MUSCLE.

B. SUPRASPINATUS MUSCLE.

C. INFRASPINATUS MUSCLE.

D. TERES MINOR MUSCLE.

E. TERES MAJOR MUSCLE.

F. LATISSIMUS DORSI MUSCLE.

G. PECTORALIS MAJOR MUSCLE.

H. SUBSCAPULARIS MUSCLE.

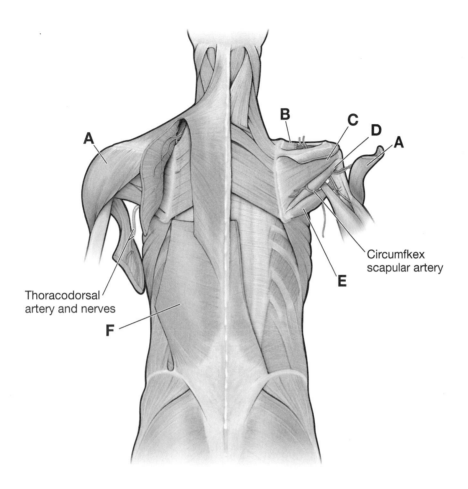

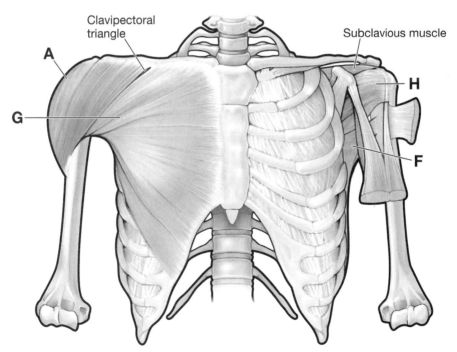

Muscles of the glenohumeral joint (posterior view and anterior view).

30e   Brachial Plexus of the Shoulder

The upper limb is innervated by the ventral rami from nerve roots C5–T1, which form a network of nerves referred to as the brachial plexus. The brachial plexus is divided into five regions consisting of roots, trucks, divisions, cords, and terminal branches.

A. DORSAL SCAPULAR NERVE.

B. SUPRASCAPULAR NERVE.

C. LONG THORACIC NERVE.

D. MEDIAL PECTORAL NERVE.

E. LATERAL PECTORAL NERVE.

F. MEDIAL CUTANEOUS NERVE OF ARM.

G. UPPER SUBSCAPULAR NERVE.

H. THORACODORSAL NERVE.

I. AXILLARY NERVE.

J. MUSCULOCUTANEOUS NERVE.

K. RADIAL NERVE.

L. MEDIAN NERVE.

M. ULNAR NERVE.

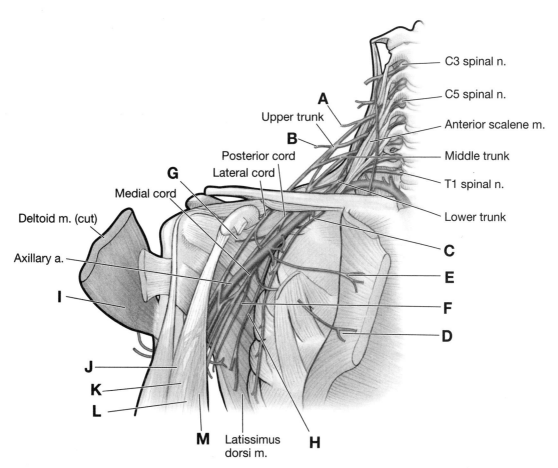

C3 spinal n.

C5 spinal n.

Anterior scalene m.

Middle trunk

T1 spinal n.

Lower trunk

Upper trunk

Posterior cord

Lateral cord

Medial cord

Deltoid m. (cut)

Axillary a.

A

B

G

I

J

K

L

M

Latissimus
dorsi m.

H

C

E

F

D

Brachial plexus and topography of the axillary artery.

30f   Posterior Division of Brachial Plexus in the Shoulder

The upper limb is innervated by the ventral rami from nerve roots C5–T1, which form a network of nerves referred to as the brachial plexus. The brachial plexus is divided into five regions consisting of roots, trucks, divisions, cords, and terminal branches.

The posterior divisions give rise to the nerves that innervate the posterior compartments of the limb (extensor muscles).

A. UPPER SUBSCAPULAR NERVE.

B. THORACODORSAL NERVE.

C. LOWER SUBSCAPULAR NERVE.

D. AXILLARY NERVE.

E. RADIAL NERVE.

F. SUPRASCAPULAR NERVE.

G. SUPRASCAPULAR ARTERY.

H. DORSAL SCAPULAR NERVE.

I. DORSAL SCAPULAR ARTERY.

J. CIRCUMFLEX SCAPULAR ARTERY.

K. DEEP ARTERY OF ARM.

L. THORACODORSAL ARTERY.

M. POSTERIOR HUMERAL CIRCUMFLEX ARTERY.

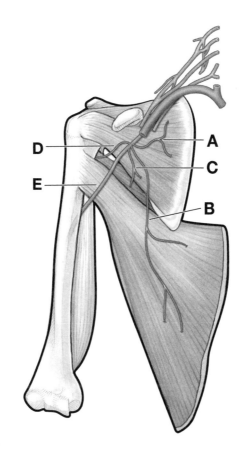

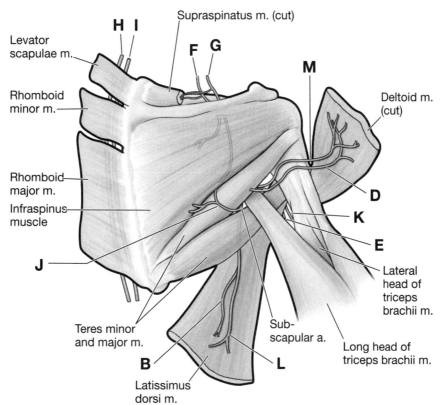

Posterior division of the brachial plexus (top, anterior view; bottom, posterior view).

30g Axillary Artery

Blood supply to both upper limbs is provided by the subclavian arteries. The right subclavian artery is a branch from the brachiocephalic artery, and the left subclavian artery is a branch from the aortic arch. The subclavian artery becomes the axillary artery as it crosses over the lateral border of the first rib. The axillary artery continues distally and becomes the brachial artery at the inferior border of the teres major muscle. The brachial artery continues distally, passing over the elbow, and becomes the ulnar and radial arteries.

A. SUBCLAVIAN ARTERY.

B. FIRST PART OF AXILLARY ARTERY.

C. SUPERIOR THORACIC ARTERY.

D. SECOND PART OF AXILLARY ARTERY.

E. THORACOACROMIAL TRUNK.

F. PECTORAL ARTERY.

G. LATERAL THORACIC ARTERY.

H. THIRD PART OF AXILLARY ARTERY.

I. POSTERIOR CIRCUMFLEX HUMERAL ARTERY.

J. CIRCUMFLEX SCAPULAR ARTERY.

K. SUBSCAPULAR ARTERY.

L. THORACODORSAL ARTERY.

M. ANTERIOR CIRCUMFLEX ARTERY.

N. BRACHIAL ARTERY.

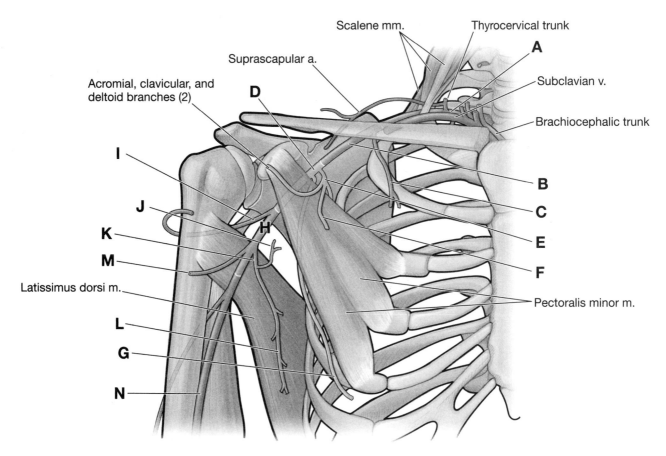

Scalene mm.

Thyrocervical trunk

Suprascapular a.

A

Acromial, clavicular, and
deltoid branches (2)

Subclavian v.

D

Brachiocephalic trunk

I

B

J

C

K

E

H

F

M

Latissimus dorsi m.

Pectoralis minor m.

L

G

N

Branches of the axillary artery.

30h   Posterior View of Shoulder Arteries and Anastomosis

If there is a blood clot blocking the subclavian artery or the artery is surgically clamped or a segment is removed, blood can bypass the blockage and reach the arm because of the rich shoulder anastomosis with the dorsal scapular, supraclavicular, and posterior humeral circumflex arteries.

A. SUBCLAVIAN ARTERY.

B. AXILLARY ARTERY.

C. SUPRASCAPULAR ARTERY.

D. DORSAL SCAPULAR ARTERY.

E. POSTERIOR HUMERAL CIRCUMFLEX ARTERY.

F. BRACHIAL ARTERY.

G. DEEP BRACHIAL ARTERY.

H. THORACODORSAL ARTERY.

I. CIRCUMFLEX SCAPULAR ARTERY.

J. SUBSCAPULAR ARTERY.

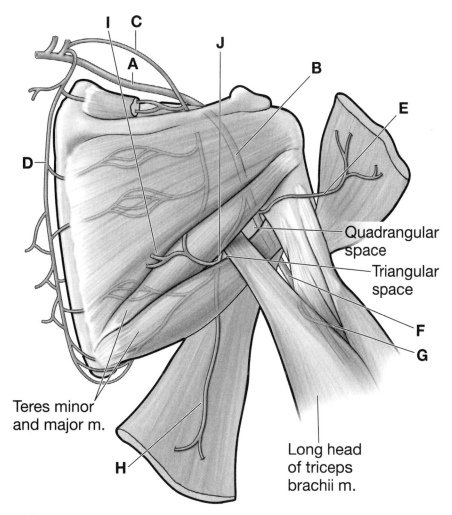

Quadrangular
space

Triangular
space

Teres minor
and major m.

Long head
of triceps
brachii m.

Posterior view of shoulder arteries.

30i   Glenohumeral Joint

The glenohumeral joint allows for a considerable amount of range of motion; more than any other joint in the body. The glenohumeral joint is a ball-and-socket synovial joint that produces a great deal of freedom, including flexion and extension, abduction and adduction, and medial and lateral rotation of the humerus. Because of the large range of motion in the glenohumeral joint it must depend upon ligaments and muscles for structural support. To minimize friction, bursae (synovial sacs) are positioned between the rotator cuff muscles and the joint capsule.

A. GLENOID CAVITY.

B. HEAD OF HUMERUS.

C. ACROMION.

D. CLAVICLE.

E. CORACOID PROCESS.

F. GREATER TUBERCLE.

G. LESSER TUBERCLE.

H. INTERTUBERCULAR GROOVE.

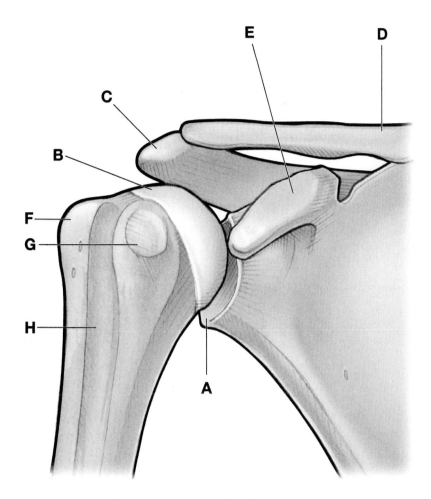

Osteology of the glenohumeral joint.

30j   Rotator Cuff Muscles

Any muscle that crosses the glenohumeral joint and produces a compressive force between the head of the humerus and the glenoid cavity will produce muscle stability. Muscle stability is best exemplified by the rotator cuff muscles, which provide support to all sides, except the inferior aspect of the glenohumeral joint.

A. SUPRASPINATUS MUSCLE.

B. INFRASPINATUS MUSCLE.

C. TERES MINOR MUSCLE.

D. SUBSCAPULARIS MUSCLE.

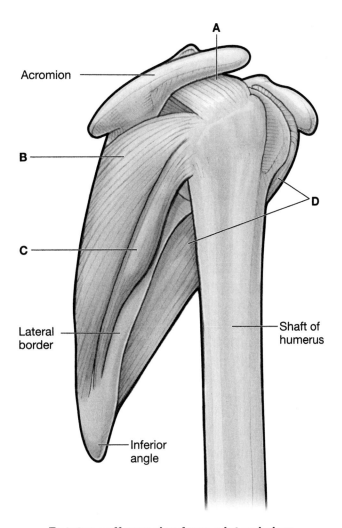

Acromion

A

B

C

Lateral
border

Inferior
angle

D

Shaft of
humerus

Rotator cuff muscles from a lateral view.

31.

Arm

31a · Brachial Muscles

The muscles of the arm are divided by their fascial compartments (anterior and posterior), and may cross one or more joints. Identifying the joints that the muscles cross and the side on which they cross can provide useful insight into the actions of these muscles.

The muscles in the anterior compartment of the arm are primarily flexors (of the shoulder or elbow or both) because of their anterior orientation. The musculocutaneous nerve (C5–C7) innervates the muscles in the anterior compartment of the arm. However, each muscle does not necessarily receive each spinal nerve level between C5 and C7.

The muscles in the posterior compartment of the arm are primarily extensors of the shoulder and elbow because of their posterior orientation. The radial nerve (C6–C8) innervates the muscle in the posterior compartment of the arm.

A. CORACOBRACHIALIS MUSCLE.

B. SHORT HEAD OF BICEPS BRACHII MUSCLE.

C. LONG HEAD OF BICEPS BRACHII MUSCLE.

D. BRACHIALIS MUSCLE.

E. SUPRASPINATUS MUSCLE.

F. SUBSCAPULARIS MUSCLE.

G. TERES MAJOR MUSCLE.

H. TERES MINOR MUSCLE.

I. INFRASPINATUS MUSCLE.

J. LONG HEAD OF TRICEPS BRACHII MUSCLE.

K. LATERAL HEAD OF TRICEPS BRACHII MUSCLE.

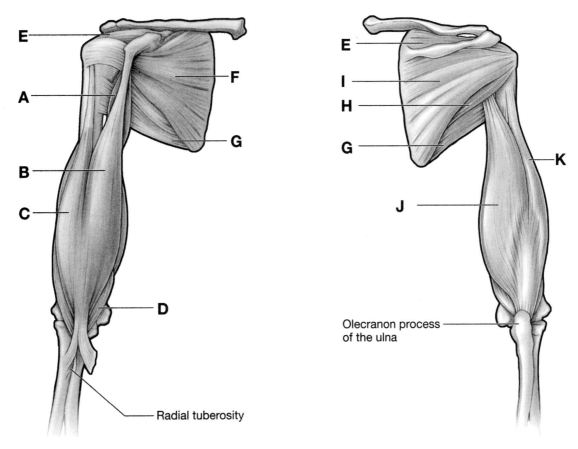

E

A

B

C

F

G

D

Radial tuberosity

E

I

H

G

K

J

Olecranon process
of the ulna

Brachial muscles (anterior and posterior views).

31b   Innervation of the Arm (Anterior Compartment)

Innervation to the anterior and posterior compartments of the arm originates from the lateral and posterior cord, giving rise to the musculocutaneous and radial nerves, respectively. Both nerves are mixed and provide motor and sensory innervation.

The musculocutaneous nerve pierces the coracobrachialis muscle, innervating it as it passes, and descends through the arm between the biceps brachii and brachialis muscles, supplying both muscles. The musculocutaneous nerve pierces the deep fascia just distal to the elbow to become the lateral cutaneous nerve of the forearm.

A. LATERAL CORD OF BRACHIAL PLEXUS.

B. MUSCULOCUTANEOUS NERVE.

C. CORACOBRACHIALIS MUSCLE.

D. BICEPS BRACHII MUSCLE.

E. BRACHIALIS MUSCLE.

F. LATERAL ANTEBRACHIAL CUTANEOUS NERVE.

G. AXILLARY ARTERY.

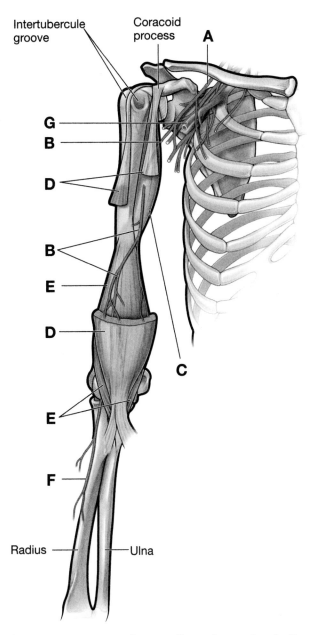

Intertubercule groove

Coracoid process

A

G

B

D

B

E

D

C

E

F

Radius   Ulna

Musculocutaneous nerve innervation of muscles in the anterior compartment of the arm.

31c   Innervation of the Arm (Posterior Compartment)

Innervation to the anterior and posterior compartments of the arm originates from the lateral and posterior cord, giving rise to the musculocutaneous and radial nerves, respectively. Both nerves are mixed and provide motor and sensory innervation.

The radial nerve descends posterior to the humerus with the deep artery of the arm, supplying motor innervation to the triceps brachii. It provides cutaneous innervation to the posterior region of the arm and forearm.

A. POSTERIOR CORD OF BRACHIAL PLEXUS.

B. RADIAL NERVE.

C. POSTERIOR CUTANEOUS NERVE OF ARM.

D. INFERIOR LATERAL CUTANEOUS NERVE OF ARM.

E. POSTERIOR CUTANEOUS NERVE OF ARM.

F. SUPERFICIAL BRANCH OF RADIAL NERVE.

G. DEEP BRANCH OF RADIAL NERVE.

H. POSTERIOR INTEROSSEOUS NERVE.

I. BRACHIALIS MUSCLE.

J. SUPINATOR MUSCLE.

K. AXILLARY ARTERY.

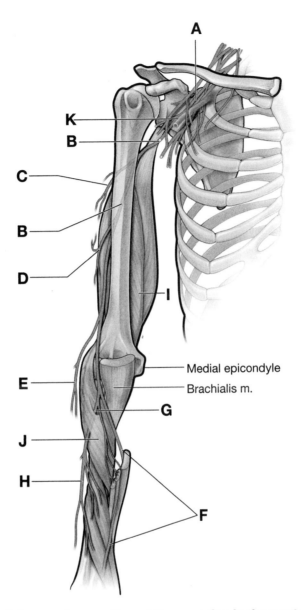

Radial nerve innervation of the muscles in the posterior compartment of arm.

| 31d | **Arteries of Arm** |

The brachial artery courses through the medial side of the anterior compartment of the arm, supplying the muscles of the anterior compartment. This is accomplished through the following arteries:

A. AXILLARY ARTERY.

B. TERES MAJOR MUSCLE.

C. DEEP ARTERY OF ARM.

D. BRACHIAL ARTERY.

E. MIDDLE COLLATERAL ARTERY.

F. SUPERIOR ULNAR COLLATERAL ARTERY.

G. INFERIOR ULNAR COLLATERAL ARTERY.

H. RADIAL COLLATERAL ARTERY.

I. RADIAL ARTERY.

J. ULNAR ARTERY.

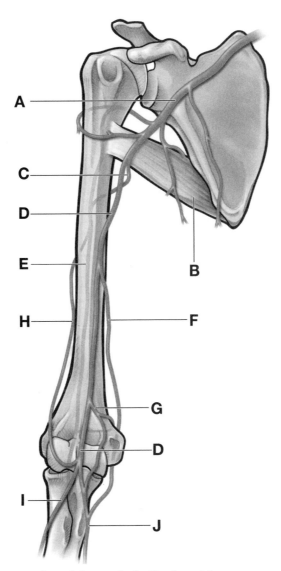

Arterial supply to the brachium.

32.

Forearm

| 32a | **Forearm Muscles (Anterior Compartment)** |

The actions produced by the muscles in the anterior compartment of the forearm depend upon which joints the muscles cross. Some muscles cross the elbow, wrist, digits, and perhaps a combination of each. The muscles in the anterior compartment of the forearm have the following similar features:

■ Common attachment. Medial epicondyle of the humerus.

■ Common innervation. Median nerve with minimal contribution from the ulnar nerve.

■ Common action. Flexion.

The muscles in the anterior compartment of the forearm are divided into three groups: superficial, intermediate, and deep.

A. BRACHIORADIALIS MUSCLE.

B. PRONATOR TERES MUSCLE.

C. FLEXOR CARPI RADIALIS MUSCLE.

D. PALMARIS LONGUS MUSCLE.

E. FLEXOR CARPI ULNARIS MUSCLE.

F. FLEXOR POLLICIS LONGUS MUSCLE.

G. FLEXOR DIGITORUM SUPERFICIALIS MUSCLE.

H. PRONATOR QUADRATUS MUSCLE.

I. FLEXOR DIGITORUM PROFUNDUS MUSCLE.

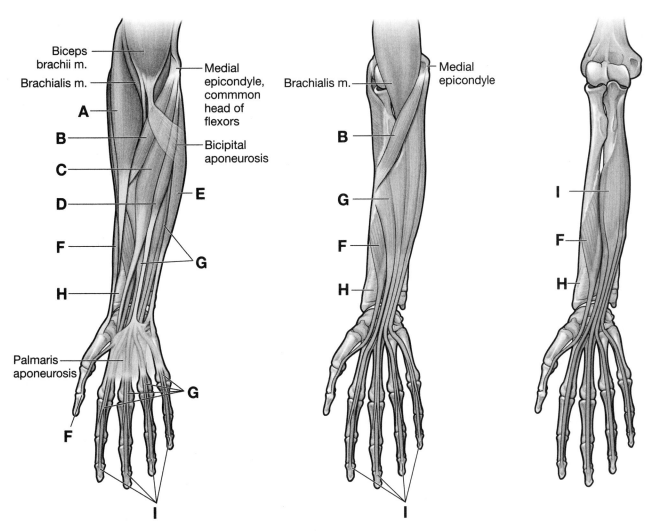

Biceps brachii m.

Brachialis m.

A

B

C

D

F

H

Palmaris aponeurosis

F

Medial epicondyle, commmon head of flexors

Bicipital aponeurosis

E

G

G

Brachialis m.

B

G

F

H

Medial epicondyle

I

F

H

I

I

Anterior forearm muscles (superficial, intermediate, and deep views).

32b   Forearm Muscles (Posterior Compartment)

The actions produced by the muscles in the posterior compartment of the forearm depend upon which joints the muscles cross. Some muscles cross the elbow, wrist, and digits, and perhaps a combination of each. The muscles in the posterior compartment of the forearm have the following similar features:

- Common attachment. Lateral epicondyle of the humerus.

- Common innervation. Radial nerve.

- Common action. Extension.

The muscles in the posterior compartment are divided into superficial and deep groups.

A. BRACHIORADIALIS MUSCLE.

B. EXTENSOR CARPI RADIALIS LONGUS MUSCLE.

C. EXTENSOR CARPI RADIALIS BREVIS MUSCLE.

D. EXTENSOR DIGITORUM MUSCLE.

E. EXTENSOR CARPI ULNARIS MUSCLE.

F. ABDUCTOR POLLICIS LONGUS MUSCLE.

G. EXTENSOR POLLICIS BREVIS MUSCLE.

H. EXTENSOR POLLICIS LONGUS MUSCLE.

I. EXTENSOR DIGITI MINIMI MUSCLE.

J. ANCONEUS MUSCLE.

K. EXTENSOR INDICIS MUSCLE.

L. SUPINATOR MUSCLE.

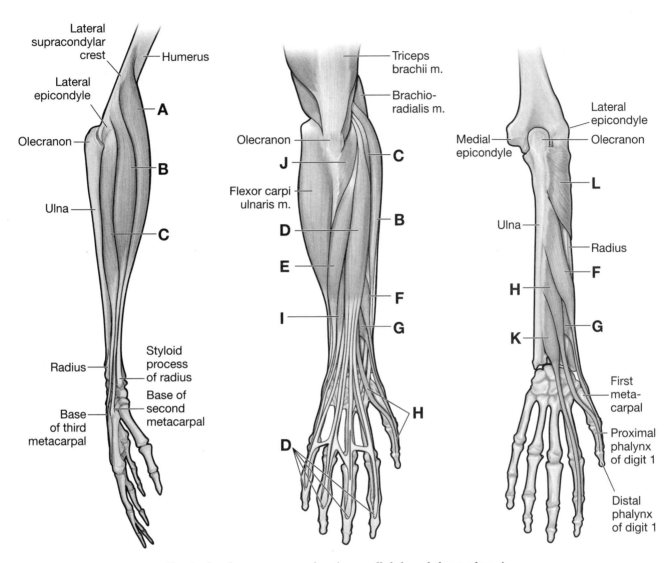

Posterior forearm muscles (superficial and deep views).

32c   Forearm Nerves—Median Nerve

The median nerve arises from the medial and lateral cords of the brachial plexus and travels with the brachial artery along the medial side of the arm. In the elbow, the median nerve courses through the cubital fossa, deep to the bicipital aponeurosis and between the two heads of the pronator teres, to enter the anterior compartment of the forearm. The median nerve continues distally to travel through the carpal tunnel to enter the hand.

A. MEDIAN NERVE.

B. ANTERIOR INTEROSSEOUS NERVE.

C. PALMAR BRANCH OF MEDIAN NERVE.

D. RECURRENT MEDIAN NERVE.

E. DIGITAL BRANCHES OF MEDIAN NERVE.

MEDIAN NERVE

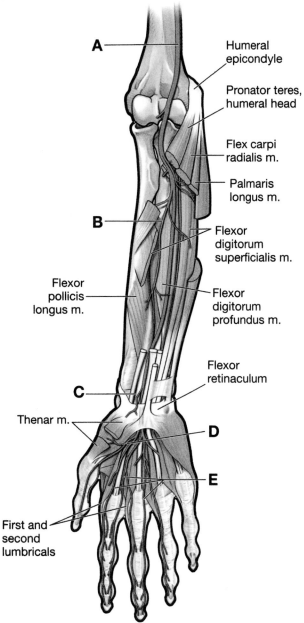

A

Humeral
epicondyle

Pronator teres,
humeral head

Flex carpi
radialis m.

Palmaris
longus m.

B

Flexor
digitorum
superficialis m.

Flexor
pollicis
longus m.

Flexor
digitorum
profundus m.

C

Flexor
retinaculum

Thenar m.

D

E

First and
second
lumbricals

Median nerve in the forearm.

32d | Forearm Nerves—Ulnar Nerve

The ulnar nerve courses posteriorly to the medial epicondyle of the humerus in the osseous groove, into the anterior compartment of the forearm between the two heads of the flexor carpi ulnaris muscle. The ulnar nerve continues through the anterior compartment of the forearm supplying only two muscles, the flexor carpi ulnaris and the ulnar half of the flexor digitorum profundus. Proximal to the wrist, the ulnar nerve gives rise to two cutaneous branches, a dorsal branch and a palmar branch, which provide cutaneous innervation to the dorsal medial side of the hand and the medial side of the palm, respectively. The ulnar nerve continues into the hand superficial to the carpal tunnel and courses through Guyon canal by the pisiform bone to enter the hand.

A. ULNAR NERVE.

B. DORSAL BRANCH OF ULNAR NERVE.

C. DIGITAL BRANCHES OF ULNAR NERVE.

ULNAR NERVE

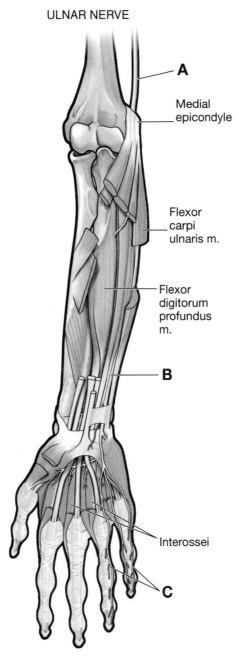

A

Medial epicondyle

Flexor carpi ulnaris m.

Flexor digitorum profundus m.

B

Interossei

C

Ulnar nerve in the forearm.

32e   Forearm Nerves—Radial Nerve

The radial nerve enters the forearm, anterior to the lateral epicondyle, and travels distally between the brachialis and the brachioradialis muscles, where it bifurcates into a deep terminal branch and a superficial terminal branch. The deep terminal branch becomes the posterior interosseous nerve, and the superficial terminal branch becomes the superficial radial nerve. The posterior cutaneous nerve of the forearm, which branches in the arm, provides sensory innervation to the posterior forearm.

A. RADIAL NERVE.

B. POSTERIOR CUTANEOUS NERVE OF FOREARM.

C. DEEP BRANCH OF RADIAL NERVE.

D. POSTERIOR INTEROSSEOUS NERVE.

E. SUPERFICIAL BRANCH OF RADIAL NERVE.

F. DIGITAL BRANCHES OF RADIAL NERVE.

RADIAL NERVE

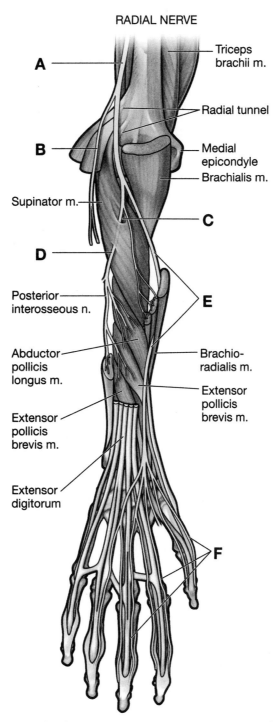

A

B

Supinator m.

C

D

Posterior
interosseous n.

Abductor
pollicis
longus m.

Extensor
pollicis
brevis m.

Extensor
digitorum

Triceps
brachii m.

Radial tunnel

Medial
epicondyle

Brachialis m.

E

Brachio-
radialis m.

Extensor
pollicis
brevis m.

F

Radial nerve in the forearm (right forearm in pronated position).

32f   Arteries of the Forearm

The brachial artery extends from the inferior border of the teres major muscle, giving rise to several branches that supply blood to the anterior and posterior compartments of the arm. The brachial artery bifurcates into the ulnar and radial arteries at the radioulnar joint. The radial and ulnar arteries and their tributaries supply blood to the anterior and posterior compartments of the forearm and extend distally into the hand.

A. BRACHIAL ARTERY.

B. SUPERIOR ULNAR COLLATERAL ARTERY.

C. INFERIOR ULNAR COLLATERAL ARTERY.

D. DEEP ARTERY OF ARM.

E. ANTERIOR ULNAR RECURRENT ARTERY.

F. POSTERIOR ULNAR RECURRENT ARTERY.

G. ULNAR ARTERY.

H. COMMON INTEROSSEOUS ARTERY.

I. ANTERIOR INTEROSSEOUS ARTERY.

J. POSTERIOR INTEROSSEOUS ARTERY.

K. INTEROSSEOUS RECURRENT ARTERY.

L. RADIAL ARTERY.

M. RADIAL RECURRENT ARTERY.

N. RADIAL COLLATERAL ARTERY.

O. MIDDLE COLLATERAL ARTERY.

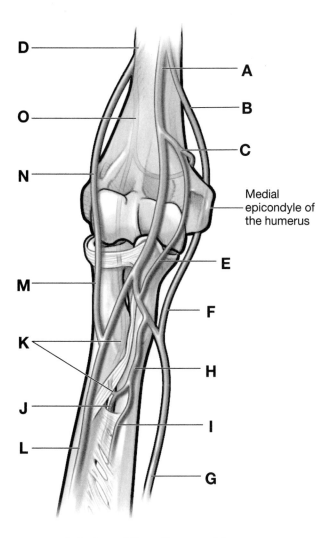

D ———

A

B

C

O ———

Medial
epicondyle of
the humerus

N ———

E

M ———

F

K ———

H

J ———

I

L ———

G

Arteries of the elbow and forearm.

33.

···

Hand

33a   Muscles of the Hand

Muscles that act on the joints of the hand can be either extrinsic (originating outside the hand) or intrinsic (originating within the hand), and they may act on a single joint or on multiple joints. The result is movement of multiple joints for activities such as functional grasping or writing.

A. ABDUCTOR POLLICIS BREVIS MUSCLE.

B. OPPONENS POLLICIS MUSCLE.

C. FLEXOR POLLICIS BREVIS MUSCLE.

D. ADDUCTOR POLLICIS MUSCLE.

E. ABDUCTOR DIGITI MINIMI MUSCLE.

F. FLEXOR DIGITI MINIMI MUSCLE.

G. OPPONENS DIGITI MINIMI MUSCLE.

H. FLEXOR DIGITORUM PROFUNDUS TENDONS.

I. LUMBRICAL MUSCLES.

J. FLEXOR DIGITORUM SUPERFICIALIS TENDONS.

K. DORSAL INTEROSSEI MUSCLES.

L. EXTENSOR DIGITORUM MUSCLE.

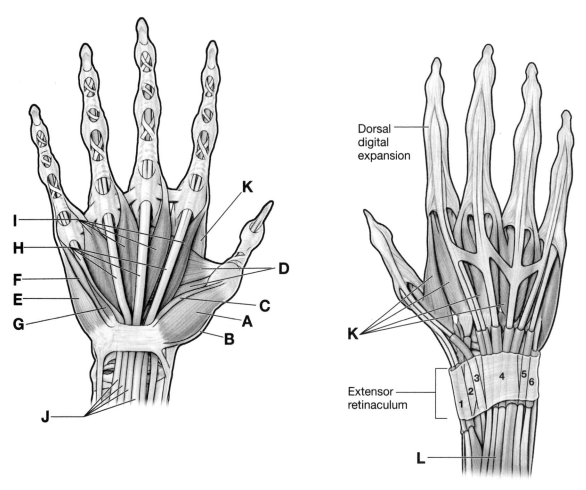

Extrinsic and intrinsic muscles of the hand.

33b   Ulnar Nerve Innervation of the Hand

The median nerve innervates lumbrical muscles 1 and 2 and the thenar muscles (excluding the deep head of the flexor pollicis brevis). The ulnar nerve provides the remaining motor innervation to the hand. The superficial radial nerve, median nerve, and ulnar nerve provide sensory innervation to the hand.

A. ULNAR NERVE.

B. DEEP BRANCH OF ULNAR NERVE.

C. SUPERFICIAL BRANCH OF ULNAR NERVE.

D. PALMAR BRANCH OF ULNAR NERVE.

E. DORSAL BRANCH OF ULNAR NERVE FROM FOREARM.

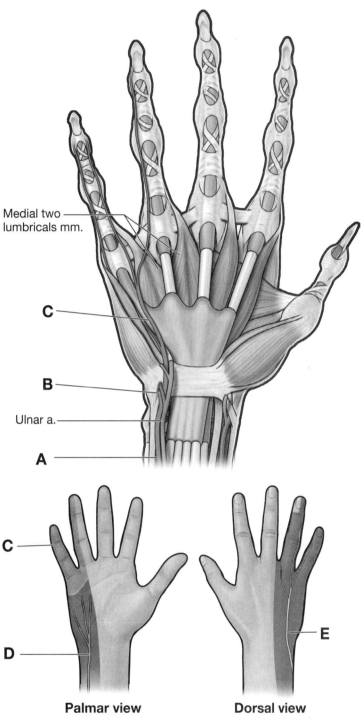

Medial two lumbricals mm.

C

B

Ulnar a.

A

Palmar view   Dorsal view
Ulnar nerve in hand.

33c	**Median Nerve Innervation of the Hand**

The median nerve innervates lumbrical muscles 1 and 2 and the thenar muscles (excluding the deep head of the flexor pollicis brevis). The ulnar nerve provides the remaining motor innervation to the hand. The superficial radial nerve, median nerve, and ulnar nerve provide sensory innervation to the hand.

A. MEDIAN NERVE.

B. PALMAR BRANCH OF MEDIAN NERVE.

C. RECURRENT BRANCH OF MEDIAN NERVE.

D. DIGITAL BRANCHES OF MEDIAN NERVE.

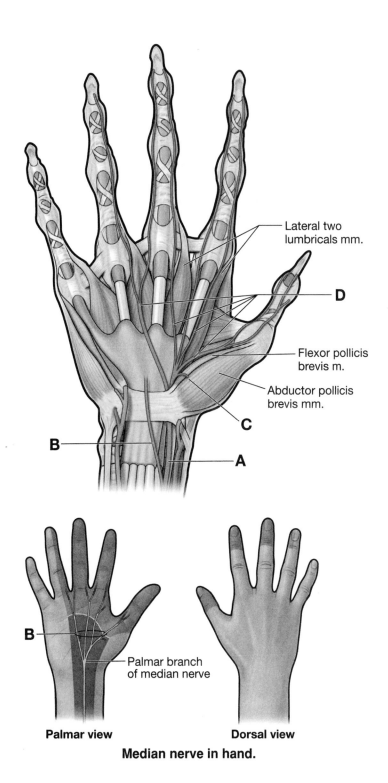

Lateral two lumbricals mm.

D

Flexor pollicis brevis m.

Abductor pollicis brevis mm.

C

B

A

B

Palmar branch of median nerve

Palmar view   **Dorsal view**

Median nerve in hand.

33d   Radial Nerve Innervation of the Hand

The superficial branch of the radial nerve enters the hand by passing superficially to the anatomical snuff box, and supplies the skin on the dorsal side of the first three digits.

The radial nerve has no motor innervation to intrinsic muscles of the hand and only innervates the extrinsic muscles that send tendons from muscles that originate in the posterior forearm to the thumb and digits.

A. SUPERFICIAL RADIAL NERVE.

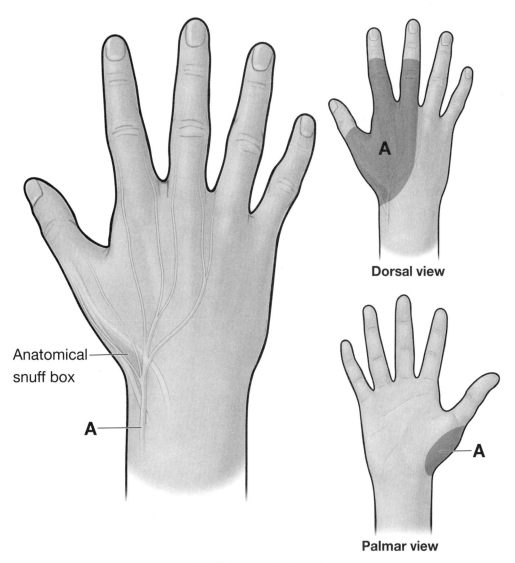

Anatomical snuff box

A

Dorsal view

Palmar view

A

Radial nerve in hand.

33e   Arteries of the Hand

The blood supply to the hand is provided by the radial and ulnar arteries, which give rise to a superficial and a deep palmar arch and to smaller tributaries as they travel distally to the tips of the fingers. The ulnar artery, with the ulnar nerve, enters the hand lateral to the pisiform, where it gives rise to the deep palmar branch and becomes the principal contributor to the superficial palmar arch. The radial artery courses through the anatomical snuff box, contributes to the dorsal carpal arch, and then travels deep into the hand, becoming the principal contributor to the deep palmar arch.

A. ULNAR ARTERY.

B. DEEP PALMAR ARTERY (ULNAR ORIGIN).

C. SUPERFICIAL PALMAR ARTERY (ULNAR ORIGIN).

D. DIGITAL ULNAR ARTERIES.

E. RADIAL ARTERY.

F. DEEP PALMAR ARTERY (RADIAL ORIGIN).

G. PRINCEPS POLLICIS ARTERY.

H. DIGITAL RADIAL ARTERIES.

I. SUPERFICIAL PALMAR ARTERY (RADIAL ORIGIN).

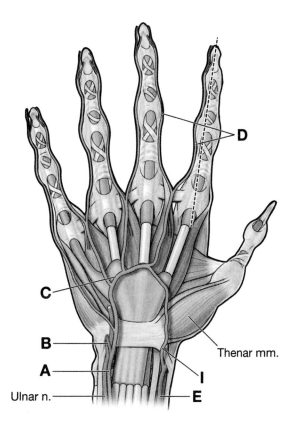

B

A

Ulnar n.

C

D

Thenar mm.

I

E

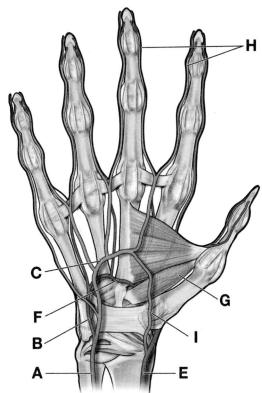

C

F

B

A

H

G

I

E

Arteries of the hand.

SECTION VII

LOWER LIMB

34.

Overview of the Lower Limb

34a   Bones of the Lower Limb

The bones of the skeleton provide a framework that serves as an attachment for soft tissues (e.g., muscles). The bony structure of the gluteal region and thigh, from proximal to distal, consists of the pelvis, femur, patella, tibia, and fibula. Synovial joints and fibrous ligaments serve to connect bones together.

A. ILIUM.

B. ISCHIUM.

C. PUBIS.

D. SACRUM.

E. FEMUR.

F. PATELLA.

G. TIBIA.

H. FIBULA.

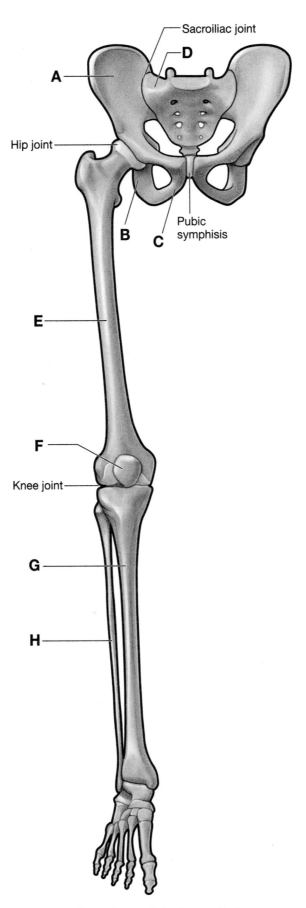

Sacroiliac joint

D

A

Hip joint

Pubic
symphisis

B

C

E

F

Knee joint

G

H

Osteology of the lower limb.

34b   Osteology of the Os Coxa and Femur

The pelvis is an irregularly shaped bone consisting of right and left pelvic bones. The pelvic bones articulate posteriorly with the sacrum, via the sacroiliac joints, and anteriorly with each other at the pubic symphysis. Each pelvic bone has three components: ilium, ischium, and pubis. The acetabulum is a large cup-shaped structure at the junction where the ilium, ischium, and pubis fuse. The acetabulum protrudes laterally for articulation with the head of the femur bone. The three bony components of the pelvis form an opening, called the obturator foramen. The femur is located in the thigh and is the longest bone of the body.

A. ILIUM.

B. ISCHIUM.

C. PUBIS.

D. GREATER TROCHANTER.

E. LESSER TROCHANTER.

F. HEAD.

G. NECK.

H. LINEA ASPERA.

I. PATELLAR FOSSA.

J. FEMORAL CONDYLES.

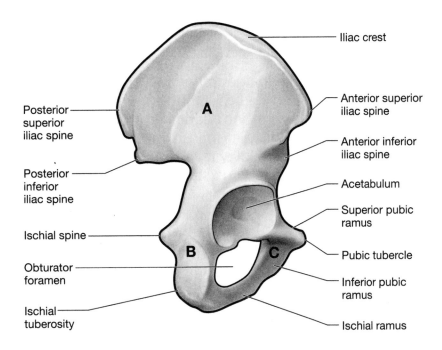

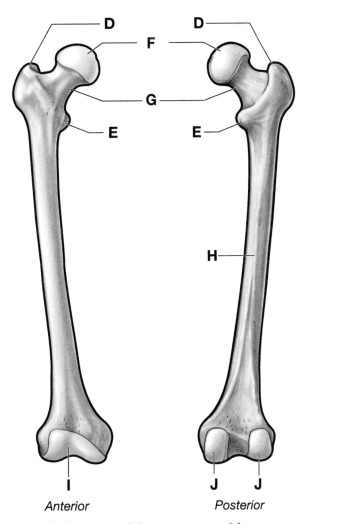

Osteology of the os coxa and femur.

34c   Osteology of the Tibia, Fibula, and Foot

The tibia is the medial and larger of the two bones of the leg. It is also the only bone that articulates with the femur at the knee joint. The fibula is lateral to the tibia and has limited involvement in weight-bearing activity. The fibular shaft is much narrower than the shaft of the tibia and is mainly enclosed by muscles. The tibia and fibula are bound together by a tough, fibrous sheath known as the interosseous membrane.

The foot is the region of the lower limb, distal to the ankle joint, and is subdivided into three sections: ankle, metatarsus, and digits. There are five digits in the foot, beginning medially with the great toe and four lateral digits. The foot has a superior surface, the dorsum, and an inferior surface, the plantar surface. The three groups of bones that comprise the foot are the 7 tarsals, 5 metatarsals, and 14 phalanges.

A. TIBIA.

B. FIBULA.

C. INTEROSSEOUS MEMBRANE.

D. TALUS.

E. CALCANEUS.

F. CUBOID.

G. NAVICULAR.

H. CUNEIFORM BONES.

I. METATARSALS.

J. PHALANGES.

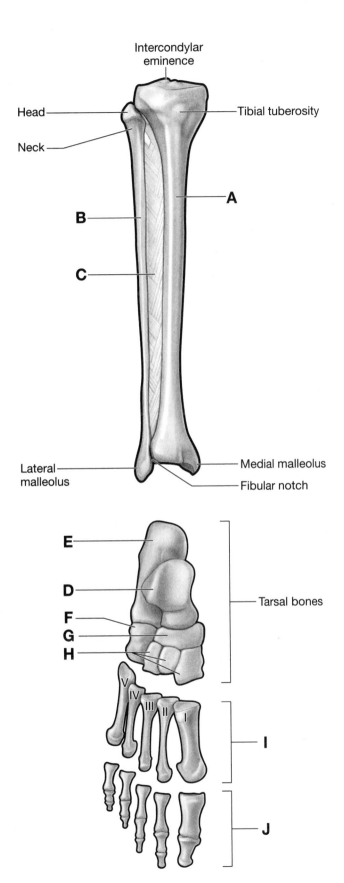

Intercondylar eminence

Head

Neck

Tibial tuberosity

A

B

C

Lateral malleolus

Medial malleolus

Fibular notch

E

D

F
G
H

Tarsal bones

V
IV
III
II
I

I

J

Osteology of the tibia, fibula, and foot.

34d   Compartments of the Thigh and Leg

Two fascial layers, known as the superficial fascia and the deep fascia, are located between the skin and bone of the lower limb. The deep fascia divides the lower limb into anterior, medial, and posterior compartments of the thigh (fascia lata) and anterior, lateral, and posterior compartments of the leg (crural fascia). Muscles are organized into these compartments and possess common attachments, innervation, and actions.

A. ANTERIOR COMPARTMENT OF THIGH. Muscles are innervated by the femoral nerve and primarily extend the knee.

B. MEDIAL COMPARTMENT OF THIGH. Muscles are innervated by the obturator nerve and primarily adduct the hip.

C. POSTERIOR COMPARTMENT OF THIGH. Muscles are innervated by the tibial nerve and primarily extend the hip and flex the knee.

D. ANTERIOR COMPARTMENT OF LEG. Muscles are innervated by the deep fibular nerve and primarily dorsiflex the ankle and extend the toes.

E. LATERAL COMPARTMENT OF LEG. Muscles are innervated by the superficial fibular nerve and primarily plantar flex and evert the ankle.

F. POSTERIOR COMPARTMENT OF LEG. Muscles are innervated by the tibial nerve and primarily plantar flex the ankle and flex the toes.

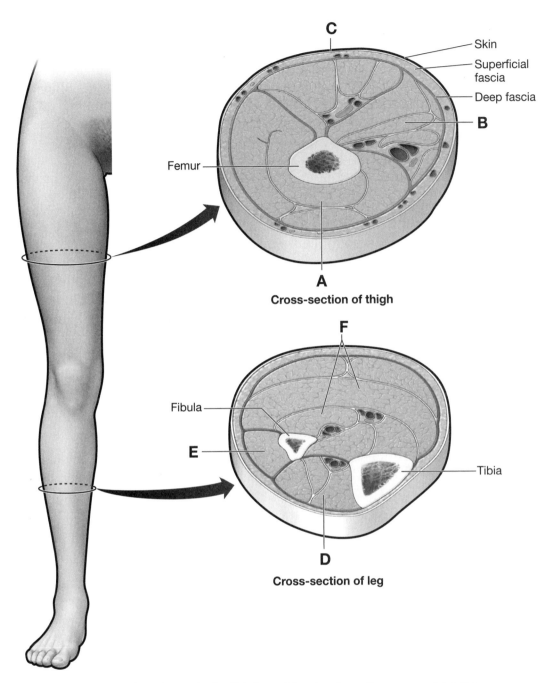

Skin

Superficial fascia

Deep fascia

Femur

A

Cross-section of thigh

Fibula

Tibia

D

Cross-section of leg

Cross-section through the thigh and leg to show the compartments.

34e Lumbosacral Plexus

The lower limb receives sensory and motor innervation from anterior rami that originate from spinal nerve levels L1–S4. The anterior rami form two networks of nerves, referred to as the lumbar plexus (L1–L4) and the sacral plexus (L4–S4), which are connected via the lumbosacral trunk (L4–L5).

Lumbar plexus. Originates from ventral rami of L1–L4. It provides motor and sensory contributions to the anterior and medial compartment of the leg as well as to the abdominal wall and pelvic areas.

Sacral plexus. Originates from ventral rami L4–S4 and is subdivided into anterior and posterior divisions. The anterior division provides the primary motor innervation to the posterior compartment of the thigh, leg, and foot. The posterior division provides the primary motor innervation to the anterior and lateral compartments of the leg.

A. SUBCOSTAL NERVE.

B. ILIOHYPOGASTRIC NERVE.

C. ILIOINGUINAL NERVE.

D. GENITOFEMORAL NERVE.

E. LATERAL FEMORAL CUTANEOUS NERVE.

F. FEMORAL NERVE.

G. OBTURATOR NERVE.

H. SUPERIOR GLUTEAL NERVE.

I. INFERIOR GLUTEAL NERVE.

J. COMMON FIBULAR NERVE.

K. TIBIAL NERVE.

L. SCIATIC NERVE.

M. POSTERIOR FEMORAL CUTANEOUS NERVE.

N. PUDENDAL NERVE.

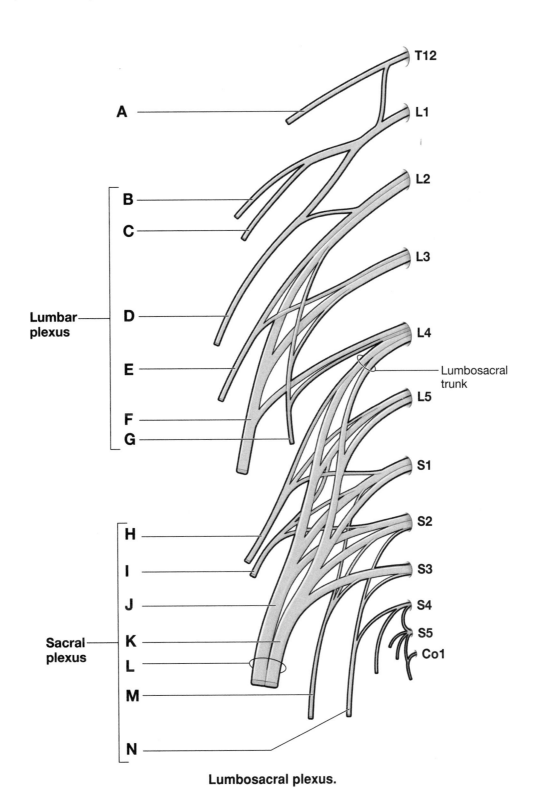

Lumbosacral plexus.

34f   Arteries of Lower Limb

The common iliac artery bifurcates into the internal and external iliac arteries. The internal iliac artery gives rise to the obturator artery and to the superior and inferior gluteal arteries. The external iliac artery becomes the femoral artery as it passes the inguinal ligament and enters the thigh. The femoral artery gives rise to the deep artery of the thigh and continues distally to become the popliteal artery behind the knee joint. The popliteal artery bifurcates into the anterior and posterior tibial arteries, which travel distally into the leg. These arteries continue into the dorsal and plantar surfaces of the foot.

A. COMMON ILIAC ARTERY.

B. EXTERNAL ILIAC ARTERY.

C. FEMORAL ARTERY.

D. DEEP ARTERY OF THE THIGH.

E. POPLITEAL ARTERY.

F. POSTERIOR TIBIAL ARTERY.

G. ANTERIOR TIBIAL ARTERY.

H. DORSAL PEDAL ARTERY.

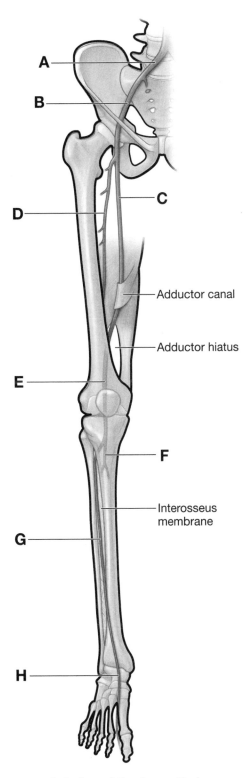

Adductor canal

Adductor hiatus

Interosseus membrane

Arteries of the lower limb.

35.

Gluteal Region and Hip

35a   Muscles of the Gluteal Region

The muscles of the gluteal region primarily act on the hip joint, producing extension, medial rotation, lateral rotation, and abduction. In addition to producing motion, the muscles of the gluteal region are important for stability of the hip joint as well as for locomotion. The deep hip rotator muscles all have several common characteristics—they are deep to the gluteal muscles, they arise from the pelvis, they share common attachments around the greater trochanter of the femur, and they laterally rotate the hip.

A. GLUTEUS MAXIMUS MUSCLE.

B. GLUTEUS MEDIUS MUSCLE.

C. TENSOR FASCIA LATA MUSCLE.

D. GLUTEUS MINIMIS MUSCLE.

E. ILIOTIBIAL TRACT.

F. PIRIFORMIS MUSCLE.

G. SUPERIOR GEMELLUS MUSCLE.

H. OBTURATOR INTERNUS MUSCLE.

I. INFERIOR GEMELLUS MUSCLE.

J. QUADRATUS FEMORIS MUSCLE.

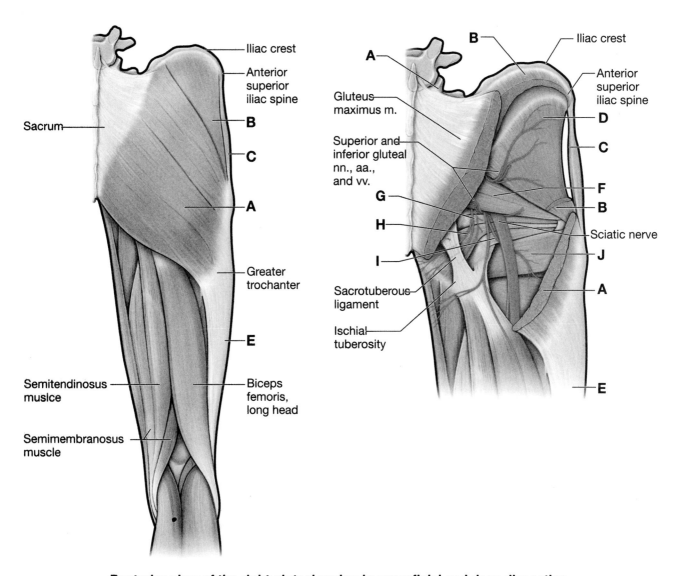

Iliac crest

Anterior superior iliac spine

B

C

A

Sacrum

Greater trochanter

E

Semitendinosus muscle

Semimembranosus muscle

Biceps femoris, long head

A

B

Iliac crest

Gluteus maximus m.

Anterior superior iliac spine

D

C

Superior and inferior gluteal nn., aa., and vv.

G

H

I

F

B

Sciatic nerve

J

Sacrotuberous ligament

Ischial tuberosity

A

E

Posterior view of the right gluteal region in superficial and deep dissection.

35b   Nerves of the Gluteal Region

The nerves in the gluteal region consist of branches from the sacral plexus (L4–S3). The superior and inferior gluteal nerves receive their names for their relationship to the piriformis muscle. The sciatic nerve is composed of two separate nerves: the tibial nerve and common fibular nerve.

A. SUPERIOR GLUTEAL NERVE.

B. INFERIOR GLUTEAL NERVE.

C. PUDENDAL NERVE.

D. NERVE TO OBTURATOR INTERNUS AND SUPERIOR GEMELLUS.

E. PERFORATING CUTANEOUS NERVE.

F. INFERIOR RECTAL NERVE.

G. POSTERIOR FEMORAL CUTANEOUS NERVE.

H. SCIATIC NERVE.

I. COMMON FIBULAR NERVE.

J. TIBIAL NERVE.

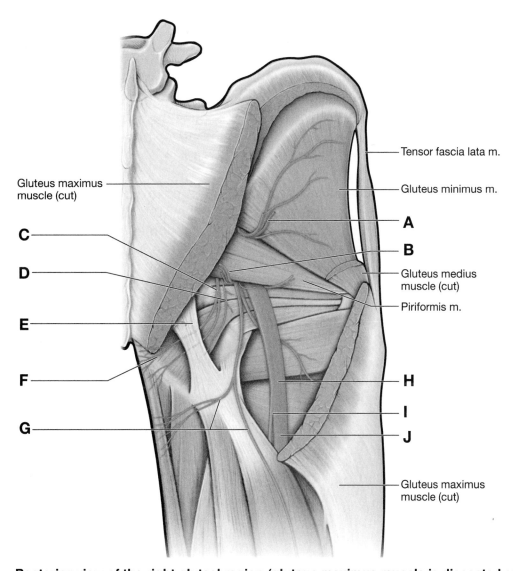

Tensor fascia lata m.

Gluteus maximus muscle (cut)

Gluteus minimus m.

A

C

B

D

Gluteus medius muscle (cut)

Piriformis m.

E

F

H

G

I

J

Gluteus maximus muscle (cut)

Posterior view of the right gluteal region (gluteus maximus muscle is dissected away).

35c   Structure of the Hip Joint

The hip joint is a synovial, ball-and-socket joint that allows for a great deal of freedom and provides support for the weight of the head, arms, and trunk. The gluteal region also contains the sacroiliac joint and the pubic symphysis, which connect the pelvic bones together as well as connecting the pelvic bones to the spine.

The articulating surface of the pelvic bone (os coxa) is a concave socket that is composed of three fused bones, the ilium, ischium, and pubis, called the acetabulum. The acetabulum is horseshoe-shaped fossa that articulates with the head of the femur. The medial circumflex femoral artery provides the principal blood supply to the hip.

The joint capsule is strong and extends like a sleeve from the acetabulum to the base of the neck of the femur. The joint capsule possesses circular fibers, which form a ring around the neck of the femur, called the zona orbicularis. The capsule contains three capsular ligaments: two anterior ligaments and one posterior ligament. The ligaments of the hip primarily become taut with extension of the hip and permit little, if any, distraction between the articulating surfaces.

A. PUBOFEMORAL LIGAMENT.

B. ILIOFEMORAL LIGAMENT.

C. ISCHIOFEMORAL LIGAMENT.

D. SACROTUBEROUS LIGAMENT.

E. SACROSPINOUS LIGAMENT.

F. SACROILIAC LIGAMENT.

G. GREATER SCIATIC FORAMEN.

H. LESSER SCIATIC FORAMEN.

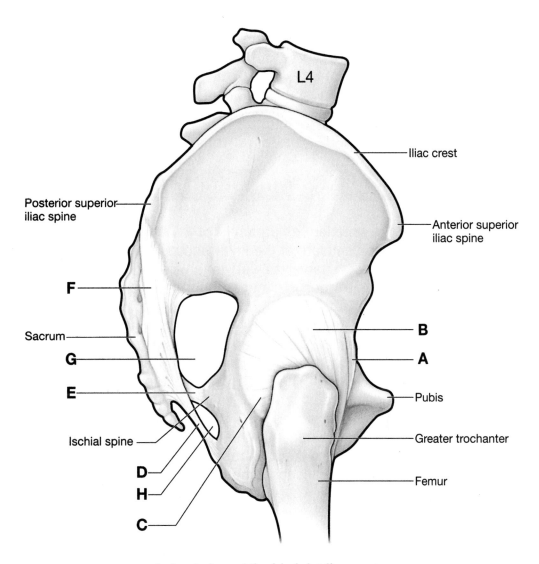

Lateral view of the hip joint ligaments.

36.

Thigh

36a   Muscles of the Anterior Thigh Compartment

The muscles in the anterior compartment of the thigh are primarily flexors of the hip or extensors of the knee because of their anterior orientation. The femoral nerve (L2–L4) innervates these muscles; however, each muscle does not necessarily receive each spinal nerve level between L2 and L4.

The quadriceps femoris muscle group consists of a four-headed muscle in the anterior compartment of the thigh and is a strong extensor muscle of the knee. There are four separate muscles in this group (rectus femoris, vastus lateralis, vastus medialis, and vastus intermedius), each with distinct origins. However, all four parts of the quadriceps femoris muscle attach to the patella, via the quadriceps tendon, and then insert onto the tibial tuberosity. The femoral nerve (L2–L4) innervates the quadriceps femoris muscle group.

A. PSOAS MAJOR MUSCLE.

B. ILIACUS MUSCLE.

C. ILIOPSOAS MUSCLE.

D. RECTUS FEMORIS MUSCLE.

E. VASTUS LATERALIS MUSCLE.

F. VASTUS MEDIALIS MUSCLE.

G. SARTORIUS MUSCLE.

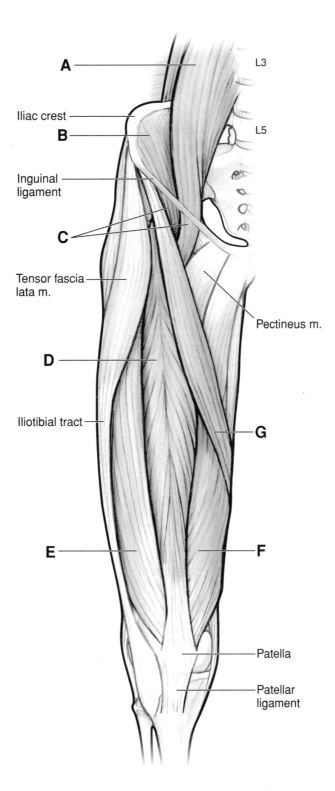

A

L3

Iliac crest

B

L5

Inguinal
ligament

C

Tensor fascia
lata m.

Pectineus m.

D

Iliotibial tract

G

E

F

Patella

Patellar
ligament

Anterior view of anterior thigh muscles (right side).

36b   Muscles of the Medial Thigh Compartment

The muscles in the medial compartment of the thigh are primarily adductors of the hip because of their medial orientation. The obturator nerve (L2–L4) innervates most of the muscles in the medial compartment of the thigh.

A. PECTINEUS MUSCLE.

B. ADDUCTOR BREVIS MUSCLE.

C. ADDUCTOR LONGUS MUSCLE.

D. ADDUCTOR MAGNUS MUSCLE.

E. GRACILIS MUSCLE.

F. ADDUCTOR HIATUS.

G. OBTURATOR EXTERNUS MUSCLE.

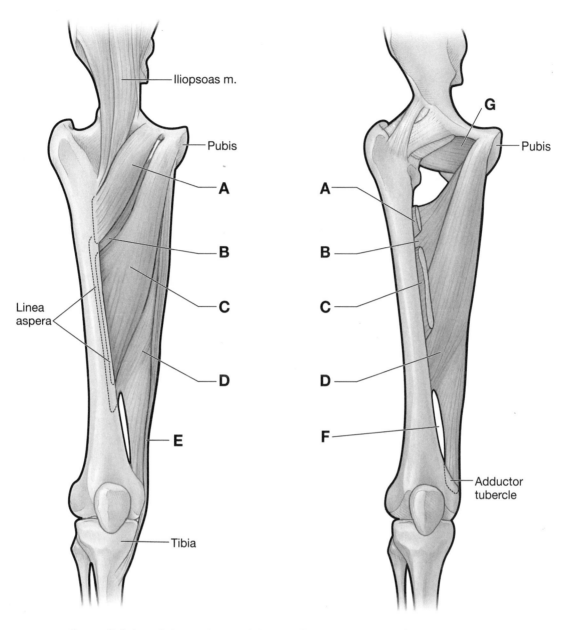

Iliopsoas m.

Pubis

A

B

C

Linea aspera

D

E

Tibia

G

Pubis

A

B

C

D

F

Adductor tubercle

Superficial and deep views of the medial compartment thigh muscles.

36c   Muscles of the Posterior Thigh Compartment

The muscles in the posterior compartment of the thigh are primarily extensors of the hip or flexors of the knee because of their posterior orientation. The tibial nerve (L4–S3) innervates the muscles in the posterior compartment of the thigh, with the exception of the short head of the biceps femoris muscles (common fibular nerve). Muscles in this compartment do not receive all the innervation levels from the tibial nerve; rather, they receive innervation from the spinal nerve level between L5 and S2.

A. SEMITENDINOSUS MUSCLE.

B. SEMIMEMBRANOSUS MUSCLE.

C. BICEPS FEMORIS MUSCLE (LONG HEAD).

D. BICEPS FEMORIS MUSCLE (SHORT HEAD).

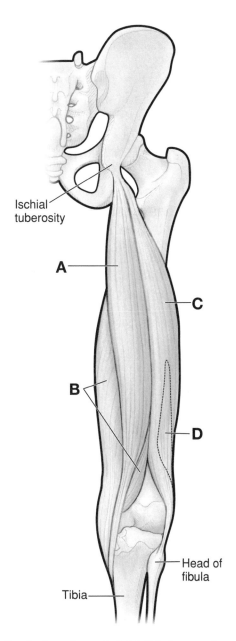

Ischial tuberosity

A

C

B

D

Head of fibula

Tibia

Posterior view of right thigh muscles.

36d   Femoral Triangle

The femoral triangle is an area in the inguinal region that is shaped like an upside-down triangle. The femoral triangle contains the femoral nerve, artery, and vein, and the lymphatics. It is bordered by the sartorius muscle, adductor longus muscle, and inguinal ligament. The femoral triangle contains the following structures from lateral to medial:

- Femoral nerve. Originates as a branch of the lumbar plexus. The femoral nerve is not contained within the femoral sheath.

- Femoral artery. Continuation of the external iliac artery. The femoral artery is located midway between the anterior superior iliac spine and the pubic symphysis.

- Femoral vein. Continues as the external iliac vein.

- Lymphatics.

Often, the acronym NAVL is used to represent the orientation of the structures of the femoral triangle.

A. INGUINAL LIGAMENT.

B. SARTORIUS MUSCLE.

C. ADDUCTOR LONGUS MUSCLE.

D. FEMORAL TRIANGLE.

E. ILIOPSOAS MUSCLE.

F. FEMORAL NERVE.

G. FEMORAL ARTERY.

H. FEMORAL VEIN.

I. FEMORAL LYMPHATICS.

J. PECTINEUS MUSCLE.

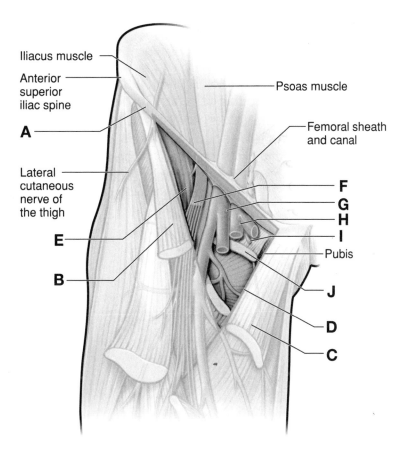

Iliacus muscle

Anterior superior iliac spine

A

Lateral cutaneous nerve of the thigh

E

B

Psoas muscle

Femoral sheath and canal

F
G
H
I

Pubis

J

D

C

Anterior view of right thigh with focus on the femoral triangle.

36e   Branches of Femoral Artery

The blood supply to the lower extremity initiates from the descending aorta, which divides into the common iliac arteries. The common iliac arteries divide into the external and internal iliac arteries. The external iliac artery passes deep to the inguinal ligament to become the femoral artery, serving as the primary blood supply to the lower limb. The internal iliac artery gives rise to the obturator artery, which also contributes to blood supply of the lower limb.

A. EXTERNAL ILIAC ARTERY.

B. FEMORAL ARTERY.

C. DEEP FEMORAL ARTERY.

D. MEDIAL FEMORAL CIRCUMFLEX ARTERY.

E. LATERAL FEMORAL CIRCUMFLEX ARTERY.

F. FEMORAL NERVE.

G. FEMORAL VEIN.

H. POPLITEAL ARTERY.

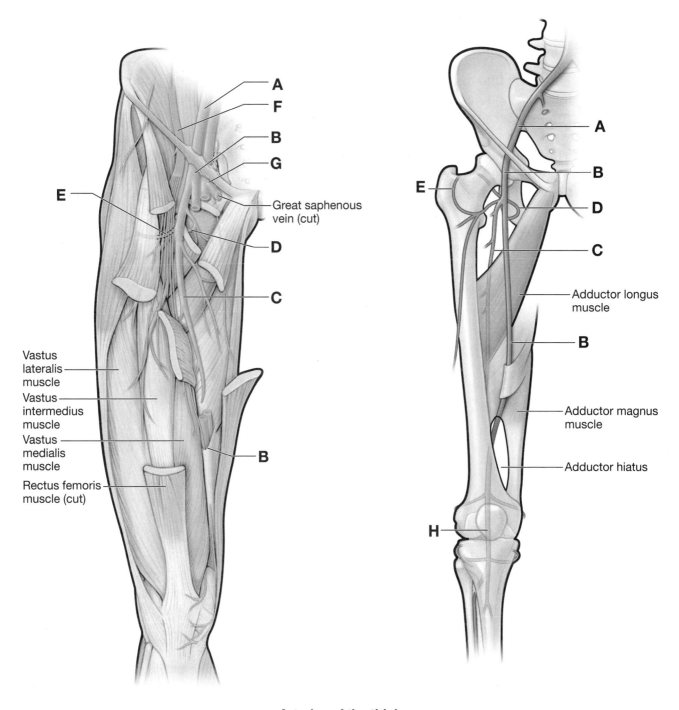

Arteries of the thigh.

36f   Ligaments of the Knee

Ligament and capsule support of the knee are critical because of the incongruence of the joint, weight bearing of the joint, and the large range of motion with flexion and extension.

The knee complex also contains ligaments inside the capsule, called the cruciate ligaments named by their location of attachment on the tibia. The knee contains two fibrocartilaginous structures, one over the medial tibial plateau, the medial meniscus, and one over the lateral tibial plateau, the lateral meniscus. Both are crescent shaped and do not complete a full circle. The menisci are also wedge shaped and, medially, are thin. However, laterally, the menisci are thicker, which increases the concavity of the articulating surface of the tibia.

A. ANTERIOR CRUCIATE LIGAMENT.

B. POSTERIOR CRUCIATE LIGAMENT.

C. LATERAL COLLATERAL LIGAMENT.

D. MEDIAL COLLATERAL LIGAMENT.

E. MEDIAL MENISCUS.

F. LATERAL MENISCUS.

G. PATELLAR LIGAMENT.

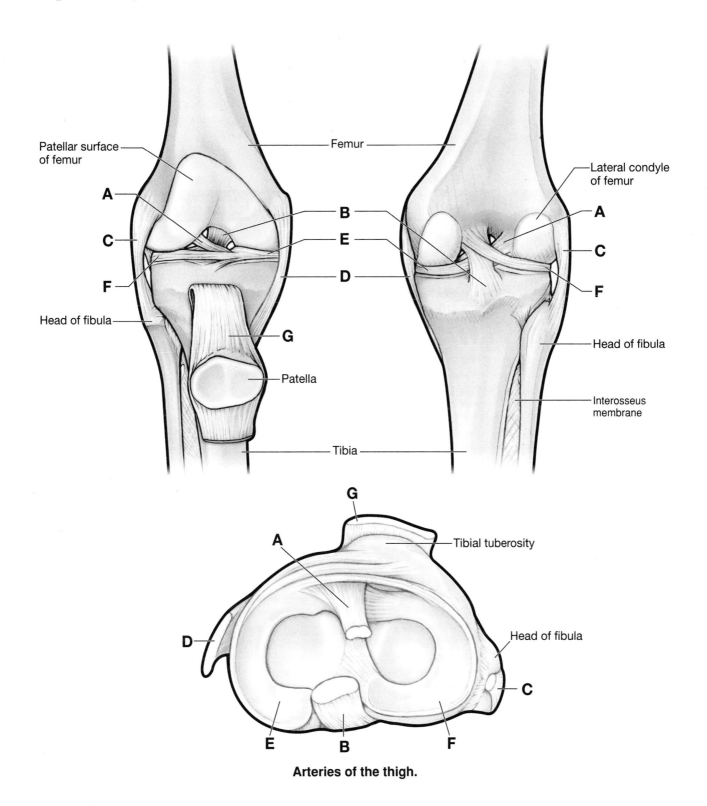

Patellar surface of femur

Femur

Lateral condyle of femur

A

B

E

D

C

F

Head of fibula

G

Patella

Head of fibula

Interosseus membrane

Tibia

G

Tibial tuberosity

A

D

Head of fibula

C

E

B

F

Arteries of the thigh.

37.

Leg

| 37a | **Muscles of the Anterior Leg Compartment** |

The muscles of the anterior compartment of the leg produce numerous actions because some muscles cross the ankle, foot, and digits, and perhaps a combination of each of these joints. The muscles in the anterior compartment of the leg are innervated by the deep fibular nerve. The primary action of the muscles is dorsiflexion.

A. TIBIALIS ANTERIOR MUSCLE.

B. EXTENSOR DIGITORUM LONGUS MUSCLE.

C. EXTENSOR HALLUCIS LONGUS MUSCLE.

D. EXTENSOR HALLUCIS BREVIS MUSCLE.

E. FIBULARIS LONGUS MUSCLE.

F. SOLEUS MUSCLE.

G. GASTROCNEMIUS MUSCLE.

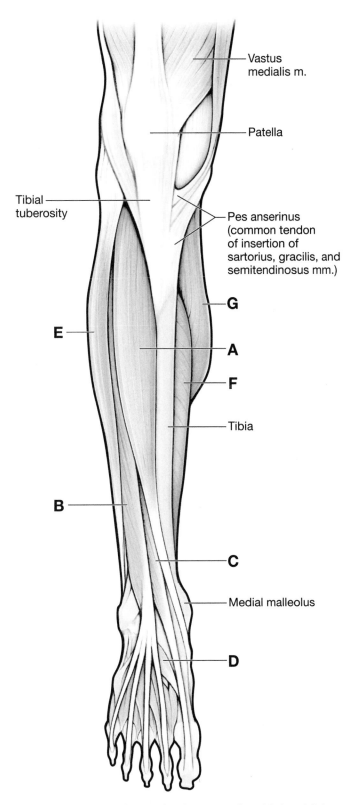

Vastus
medialis m.

Patella

Tibial
tuberosity

Pes anserinus
(common tendon
of insertion of
sartorius, gracilis, and
semitendinosus mm.)

G

E

A

F

Tibia

B

C

Medial malleolus

D

Anterior view of anterior leg muscles (right side).

37b Muscles of the Lateral Leg Compartment

The muscles of the lateral compartment of the leg produce numerous actions because some muscles cross the ankle, foot, and digits and perhaps a combination of each of these joints.

The muscles in the lateral compartment of the leg are innervated by the superficial fibular nerve. The primary action of the muscles is dorsiflexion.

A. FIBULARIS LONGUS MUSCLE.

B. FIBULARIS BREVIS MUSCLE.

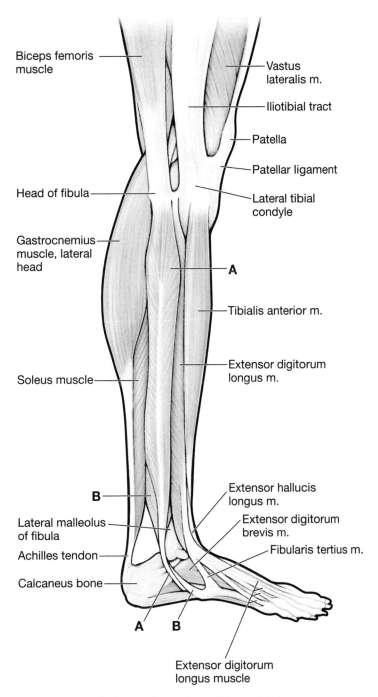

Biceps femoris muscle

Vastus lateralis m.

Iliotibial tract

Patella

Patellar ligament

Head of fibula

Lateral tibial condyle

Gastrocnemius muscle, lateral head

A

Tibialis anterior m.

Extensor digitorum longus m.

Soleus muscle

B

Extensor hallucis longus m.

Lateral malleolus of fibula

Extensor digitorum brevis m.

Fibularis tertius m.

Achilles tendon

Calcaneus bone

A   B

Extensor digitorum longus muscle

Lateral view of leg (right side).

37c   Muscles of the Posterior Leg Compartment (Superficial)

The muscles of the posterior compartment of the leg produce numerous actions because some muscles cross the ankle, foot, and digits and perhaps a combination of each of these joints. The muscles in the posterior compartment of the leg are divided into a superficial group and a deep group. Each muscle is innervated by the tibial nerve and plantar flex the ankle.

A. GASTROCNEMIUS MUSCLE.

B. SOLEUS MUSCLE.

C. ACHILLES TENDON.

D. FLEXOR DIGITORUM LONGUS MUSCLE.

E. FLEXOR HALLUCIS LONGUS MUSCLE.

F. TIBIALIS POSTERIOR MUSCLE.

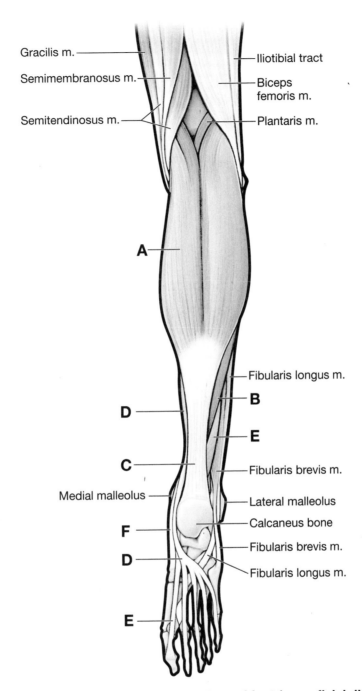

Gracilis m.

Semimembranosus m.

Semitendinosus m.

Iliotibial tract

Biceps femoris m.

Plantaris m.

A

Fibularis longus m.

B

D

E

C

Fibularis brevis m.

Medial malleolus

Lateral malleolus

Calcaneus bone

F

Fibularis brevis m.

D

Fibularis longus m.

E

Posterior view of right leg and plantar surface of foot (superficial dissection).

37d   Muscles of the Posterior Leg Compartment (Deep)

The muscles of the posterior compartment of the leg produce numerous actions because some muscles cross the ankle, foot, and digits and perhaps a combination of each of these joints. The muscles in the posterior compartment of the leg are divided into a superficial group and a deep group. Each muscle is innervated by the tibial nerve and plantar flex the ankle.

A. GASTROCNEMIUS MUSCLE.

B. SOLEUS MUSCLE.

C. PLANTARIS MUSCLE.

D. POPLITEUS MUSCLE.

E. FLEXOR DIGITORUM LONGUS MUSCLE.

F. FLEXOR HALLUCIS LONGUS MUSCLE.

G. TIBIALIS POSTERIOR MUSCLE.

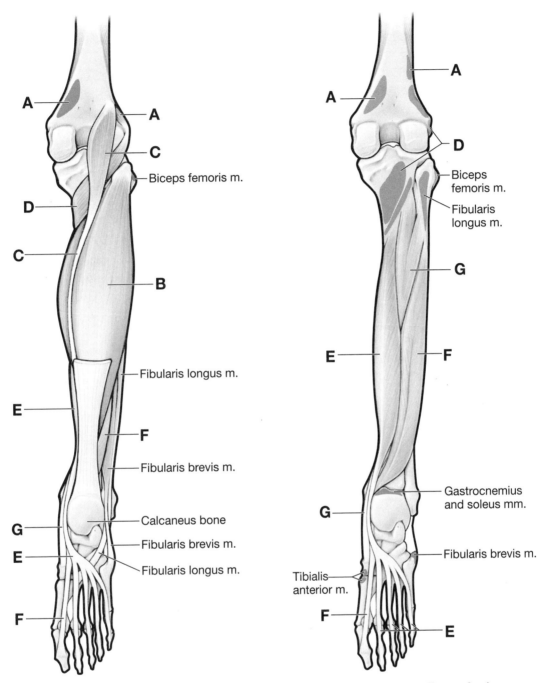

A

A

C

Biceps femoris m.

D

C

B

Fibularis longus m.

E

F

Fibularis brevis m.

Calcaneus bone

G

Fibularis brevis m.

E

Fibularis longus m.

F

A

A

D

Biceps
femoris m.

Fibularis
longus m.

G

E

F

Gastrocnemius
and soleus mm.

G

Fibularis brevis m.

Tibialis
anterior m.

F

E

Posterior view of right leg and plantar surface of foot (deep dissection).

37e   Nerves and Arteries of the Leg

The tibial nerve arises from the anterior division of the sacral plexus (L4–S3), descends through the popliteal fossa, and courses deep to the soleus muscle to innervate the superficial and deep group of muscles in the posterior compartment of the leg. The tibial nerve descends the posterior region of the leg and enters the foot inferior to the medial malleolus to innervate the plantar surface of the foot. The tibial nerve has muscular and sensory branches.

The popliteal artery is a continuation of the femoral artery. It courses through the popliteal fossa on the posterior side of the knee and bifurcates into the anterior and posterior tibial arteries at the inferior border of the popliteus muscle. The anterior and posterior tibial arteries supply blood to the leg and foot.

A. TIBIAL NERVE.

B. MEDIAL SURAL NERVE.

C. COMMON FIBULAR NERVE.

D. LATERAL SURAL NERVE.

E. SUPERFICIAL FIBULAR NERVE.

F. DEEP FIBULAR NERVE.

G. POPLITEAL ARTERY.

H. POSTERIOR TIBIAL ARTERY.

I. ANTERIOR TIBIAL ARTERY.

J. FIBULAR ARTERY.

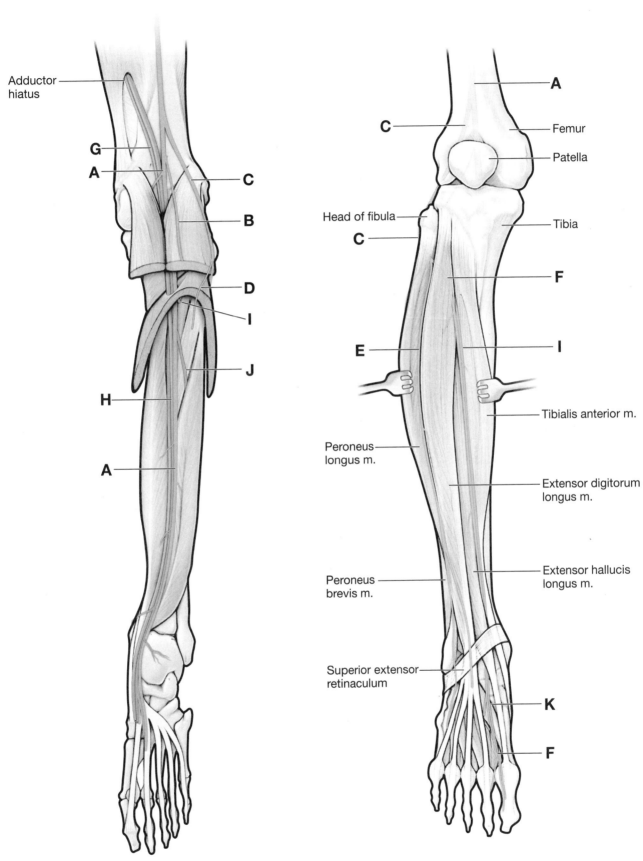

Adductor hiatus

G

A

C

B

D

I

J

H

A

A

Femur

C

Patella

Head of fibula

C

Tibia

F

E

I

Tibialis anterior m.

Peroneus longus m.

Extensor digitorum longus m.

Peroneus brevis m.

Extensor hallucis longus m.

Superior extensor retinaculum

K

F

Posterior view of right leg and plantar surface of foot (nerves and arteries).

38.

Foot

38a | Muscles of the Foot

Muscles that act on the joints of the foot can either be extrinsic (originating outside the foot) or intrinsic (originating within the foot), and they may act on a single joint or multiple joints. The result is movement of multiple joints used to accommodate uneven surfaces or for activities such as running or jumping. The medial plantar or lateral plantar nerves originate from the tibial nerve and innervate the plantar muscles of the foot. The deep fibular nerve innervates the muscles on the dorsal side. These foot muscles are divided into four layers on the plantar surface and one group on the dorsal surface.

A. FLEXOR DIGITORUM BREVIS MUSCLE.

B. ABDUCTOR DIGITI MINIMI MUSCLE.

C. ABDUCTOR HALLUCIS MUSCLE.

D. LUMBRICAL MUSCLES.

E. QUADRATUS PLANTAE MUSCLE.

F. FLEXOR DIGITI MINIMI MUSCLE.

G. FLEXOR HALLUCIS BREVIS MUSCLE.

H. ADDUCTOR HALLUCIS MUSCLE.

I. PLANTAR INTEROSSEI MUSCLES.

J. DORSAL INTEROSSEI MUSCLES.

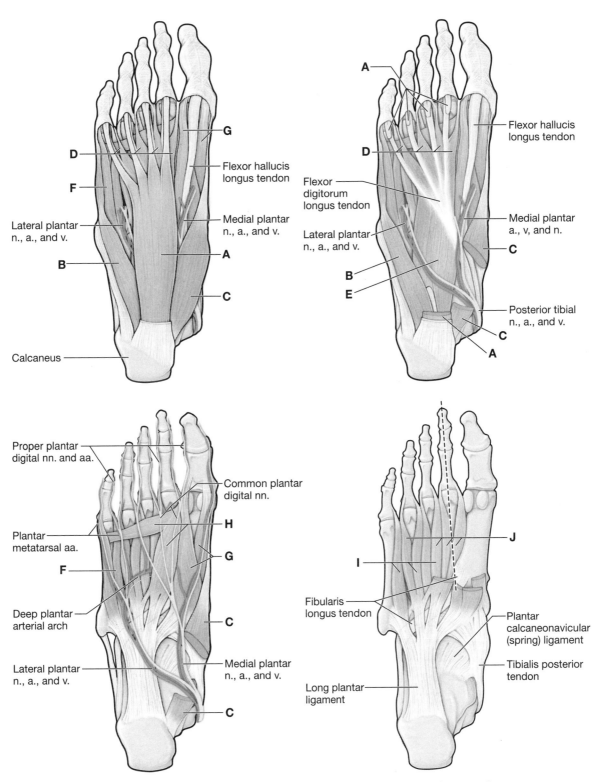

G

D

F

Flexor hallucis longus tendon

Medial plantar n., a., and v.

Lateral plantar n., a., and v.

B

A

C

Calcaneus

A

D

Flexor hallucis longus tendon

Flexor digitorum longus tendon

Lateral plantar n., a., and v.

Medial plantar a., v, and n.

C

B

E

Posterior tibial n., a., and v.

C

A

Proper plantar digital nn. and aa.

Common plantar digital nn.

Plantar metatarsal aa.

H

G

F

Deep plantar arterial arch

C

Lateral plantar n., a., and v.

Medial plantar n., a., and v.

C

J

I

Fibularis longus tendon

Plantar calcaneonavicular (spring) ligament

Long plantar ligament

Tibialis posterior tendon

Plantar surface of the foot showing four layers of intrinsic muscles.

38b   Nerves and Arteries of the Foot

The tibial nerve enters the foot inferior to the medial malleolus through the tarsal tunnel, giving rise to the medial calcaneal branch (sensory). The nerve then bifurcates into the medial and lateral plantar nerves to supply motor and sensory innervation to the plantar surface of the foot. The deep fibular nerve supplies motor innervation to the dorsum of the foot as well as sensory innervation to a small area between the first and second digits.

Sensory innervation to the medial and lateral sides of the foot is provided by the saphenous nerve (from the femoral nerve) and the sural nerve (from the tibial nerve).

A. TIBIAL NERVE.

B. MEDIAL PLANTAR NERVE.

C. LATERAL PLANTAR NERVE.

D. SUPERFICIAL FIBULAR NERVE.

E. DEEP FIBULAR NERVE.

F. SURAL NERVE.

G. SAPHENOUS NERVE.

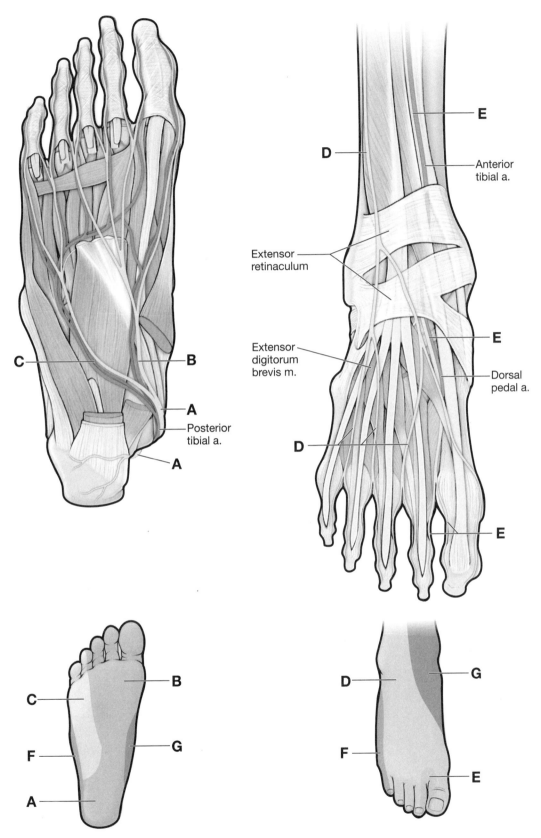

Neurovascular supply of the right foot; plantar surface (left side) and dorsal surface (right side).

E

D

Anterior
tibial a.

Extensor
retinaculum

E

Extensor
digitorum
brevis m.

Dorsal
pedal a.

D

E

C

B

A

Posterior
tibial a.

A

B

C

F

G

A

D

G

F

E

Index

Note: Page numbers followed by f indicate figure.